The Probable Universe

An owner's guide to quantum physics

Dr. M. Y. Han

TAB BOOKS

Blue Ridge Summit, PA

FIRST EDITION
FIRST PRINTING

© 1993 by **TAB Books**.
TAB Books is a division of McGraw-Hill, Inc.

Library of Congress Cataloging-in-Publication Data

Han, M. Y.
    The probable universe : an owner's guide to quantum physics / by M.Y. Han
      p. cm.
    Includes bibliographical references and index.
    ISBN 0-8306-4191-2   ISBN 0-8306-4192-0 (pbk.)
    1. Quantum theory. I. Title.
QC174.12.H358   1992
530.1′2—dc20   92-15530
                                                  CIP

TAB Books offers software for sale. For information and a catalog, please contact TAB Software Department, Blue Ridge Summit, PA 17294-0850.

Acquisitions Editor: Roland S. Phelps
Editor: John Rhea
Managing Editor: Susan Wahlman
Director of Production: Katherine G. Brown
Book Design: Jaclyn J. Boone
Cover Design: Holberg Design, York, Pa.

Contents

Acknowledgments

This book and my previous work, *The Secret Life of Quanta*, have both benefited greatly from the steady guidance, beginning to end, of my editor, Roland Phelps. For this guidance and his many helpful suggestions toward improving the quality of both of my books, I am grateful.

To the memory of my mother

Most of the fundamental ideas of science are essentially simple, and may, as a rule, be expressed in a language comprehensible to everyone.

Albert Einstein

I think I can safely say that nobody understands quantum mechanics.

Richard Feynman

＊

Prologue
Quantum physics and new technologies

In the early 1980s, a word began to appear for the first time in the names of some of the newest developments at the cutting edge of technology—the word was "tunneling." Not that it was a new word in the literal sense, but its wide and prominent use in technology was new. First came the invention in 1981 of a new breed of microscope based on a revolutionary new technique, the so-called *scanning tunneling microscope*, which for the first time provided us with stunning images of individual molecules and atoms. Then came the development, in the mid-1980s, of a radically different breed of transistors, microelectronic switches so minuscule that up to a few billion of them could fit into a sliver of silicon about the size of a thumbnail. These new transistors came to be called quantum tunneling transistors. Suddenly one of the least-known jargons from the arcane world of quantum physics—tunneling—had burst upon the front pages of technological news.

The scanning tunneling microscope was invented by Gerd Binnig and Heinrich Rohrer of Switzerland. They shared half of the 1986 Nobel prize in physics for their revolutionary achievement. A scanning tunneling microscope, or *STM*, operates with a needle-sharp tungsten tip as its probe. The tip is ground so sharp that the apex of it consists of only a few atoms. Scanning laterally across the surface of a sample in a raster-like pattern, an STM forms a computer-assisted image by recording minute movements of its probe as the tip follows the contour of the atomic landscape. STMs have produced riveting images: a precise geometrical pattern of atoms and molecules of the surface of solids; the individual atomic view of semiconductors making up an integrated circuit; and the detailed lattice structures of some of the recently discovered high-temperature superconductors. On a slightly larger scale but no less awe-inspiring, an image of an unaltered DNA molecule, looping and crossing over itself, magnified some million times by an STM, has provided scientists with their first view of the famous double helix structure (FIG. P-1). Figure P-2 shows the same image after computer enhancing.

Lawrence Berkeley Laboratory

P-I *An STM micrograph of a DNA molecule.*

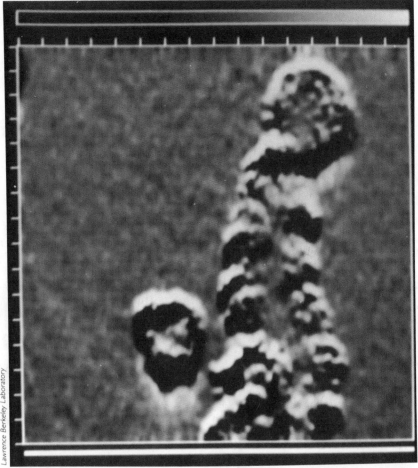

Lawrence Berkeley Laboratory

P-2 *A computer-enhanced version of the STM micrograph in FIG. P-1.*

Quantum tunneling transistors represent the latest and most advanced concepts in the design of semiconductor switching devices. Still in the experimental stages of development, these new transistors can be fabricated on a scale hundreds and even thousands of times smaller than the smallest commercially available transistors of today. Considering that some advanced commodity chips already contain as many as 10 million transistors, the enormous potency of the much-reduced dimensions of quantum tunneling transistors becomes quite evident; a much faster and more powerful microchip soon will turn into a reality such dreams as a notebook supercomputer. Many scientists believe that the development of these new transistors represents the threshold of yet another profound transition in technology, one no less revolutionary and far-reaching than the invention of the original transistors more than 40 years ago.

Tunneling, a physical process called the quantum tunneling effect, refers to one of those absolutely weird and totally befuddling phenomena in the microcosm that is the quantum world—tiny specks of matter, be they electrons or protons, can sometimes pull off the seemingly impossible trick of going from one side of an impenetrable barrier to the other side without ever actually having gone through the barrier—sort of like some supernatural shadowy figure of a dancing fog coming right through a solid wall. Impossible by all the known rules of Newtonian physics, nothing like it can ever take place in our human-sized world. The name "tunneling" reflects, in fact, a rather desperate choice; no appropriate word to describe it exists. That such a tunneling process does take place is a direct consequence—and living proof—of one of the deepest mysteries of the quantum world: the riddle of particle-wave duality. An object behaves at times as a pinpoint particle and, at other times, just as convincingly, as a spread-out wave.

As dramatic as it is, the emergence of a technology based on the quantum tunneling effect—the most direct technological application of the basic tenets of quantum physics to date—represents only the latest chapter in the history of one of the greatest developments of our time, the sprouting of sophisticated technologies from the realm of quantum mechanics. Founded during the first three decades of the twentieth century, this esoteric branch of physics has had a most profound and far-reaching impact not only on our basic understanding of the universe, but also on our technological civilization. On the one hand, it has provided the basis for our understanding of the structure and properties of atoms, and through it, the workings of molecules—bulk matter as well as biomolecules—and, by extension, what is essential about the world of atomic nuclei, elementary particles, and the fundamental forces of nature. On the other hand, the wealth of new knowledge thus gained has opened up a rapid-fire development of advanced technologies and dazzling new devices: semiconductors, microchips, microcomputers, lasers, optical fibers, and superconductors, to name a few. By the end of the eighties, the products and services springing from these new technologies accounted for up to a fourth of the gross national product of the United States, and, by some estimates, they might contribute as much as half of the combined gross national products of the industrialized world by the end of this century.

An interesting twist about all this advancement is that despite such paramount importance, very little is known of quantum mechanics outside the halls of the physics profession. Among undergraduate students in general I estimate, extrapolating from my own teaching experience, the fraction of students with some exposure to quantum mechanics to be less than one tenth of one percent—and even this might be an overestimate. Other than those majoring in physics, a group fast approaching the status of an endangered species, only a handful of the truly curious—and courageous—ever come into any substantive contact with the subject. Such extreme obscurity of quantum mechanics is, however, not without some justification. To put it bluntly, the subject easily qualifies as one of the most difficult in all of physics. The difficulties occur on at least two levels: not only is quantum mechanics mathematically complex and cumbersome—which is bad enough—but also, and even more to the point, it is philosophically weird, conceptually contradictory, and often goes directly against our common sense and intuition. Even among professionals, it is one thing to become proficient in mathematical manipulations of quantum mechanics, but quite another matter to come to terms with its conceptual ramifications.

Something that is hard to understand is so much more difficult to explain, a point well documented by an acute paucity of popular titles on the subject matter of quantum mechanics. A casual perusal of science books at an average-sized public library, for example, might turn up some 500 titles in the category of general physics. A vast majority of these, no less than 400 titles, deals with the physics of the universe at large, the legacies of Albert Einstein. The physics of space and time, the theory of relativity, astronomy, cosmology, the Big Bang, and the like represent the subject areas most written about perhaps in all science. Of the remaining titles, you might find about 80 or so dealing with the world of elementary particles, the worlds of atomic nuclei, protons, and neutrons, down to the basic building blocks of all matter, leptons and quarks. In the end you'll find only about a dozen titles explaining quantum mechanics. And so it is that almost a century after the original inception of the idea of a photon, the quantum of light, quantum mechanics remain largely hidden from those who draw the most benefit from it.

With the aim of partially remedying this situation, I have previously published a book titled *The Secret Life of Quanta* (TAB, 1990), the main thrust of which was to bring out the intimate connection between quantum physics and many of today's high-tech miracles. In that book, I went straight to the heart of atomic physics, the systematic patterns of electron configurations that define the structure of all atoms. Incorporating the further knowledge of how atoms form molecules and solids, the book provided a sufficient basis in which to introduce the physical principles behind today's technological miracles, including semiconductors, computer chips, microprocessors, lasers, optical fibers, X-ray lasers, superconductors, and magnetic resonance imaging, as well as the proposed superconducting supercolliders. Not only did I refrain from introducing even a single mathematical formula in the book (a tradition I am continuing in the present volume as well) but I also completely circumnavigated the other central difficulty of quantum physics: the paradoxical and seemingly impossible nature of its conceptual and philosophical foundations. The whole purpose of the book was, after all, to enable those not well versed in quantum physics to grasp and understand the physical principles behind today's technological marvels as efficiently and painlessly as possible.

With the advent of the scanning tunneling microscope and the development toward the eventual mass production of quantum tunneling transistors, however, we can no longer afford the luxury of circumnavigating the mysterious ways of quantum physics. The latest technology—what might be referred to as quantum tunneling technology—is singularly unique in that it represents the most direct application of the most baffling quantum reality into technological products. Now the need has clearly arisen for us to take a hard look into the soul of quantum physics and come to some understanding of the strange and baffling ways in which nature behaves in the microcosm that is the quantum world.

The quantum world is different from anything we have known previously. A concept as basic as a location of an object, a thing occupying one fixed position at one moment of time, is denied at the outset in the quantum world. A thing is not here and later there, but it is here and there at the same time with different chances for being here and being there. More than anything else, it is the rule of chances, the rule of probabilities, that separates the world of quantum from our own. The fixed position of an object is replaced by a rule of probabilities for a range of possible positions, and the speed of an object is replaced by a control under a probabilistic distribution in the range of possible speeds. The very definition of the physical world surrounding us can be specified only up to the rules of probabilities that define the possible range of realities. The quantum world is thus a microcosm of probabilities, hence the title of this book, *The Probable Universe*. This revolutionary revelation that physical reality can only be defined up to the rules of probabilities is perhaps the most difficult concept in all of quantum physics.

And therein lies the purpose of this present volume; it is intended as an introduction to the central conceptual foundations of quantum mechanics and to the phenomena of quantum tunneling and its application to the latest developments in today's technology. The book first treats the three founding principles of quantum mechanics, which together form its central mystery: wave-particle duality, the probabilistic nature of physical reality, and the resulting uncertainties inherent in all physical measurements. Next comes a discussion of the phenomenon of quantum tunneling and four examples of its technological applications: the alpha-particle emission in a class of nuclear radioactivities, a superconducting switching device called the Josephson junction switch, the scanning tunneling microscope, and the development of quantum tunneling transistors. In many fields of study, the difference between a specialist and a nonspecialist, it has been often said, boils down to the difference between knowing and not knowing what it is that we don't know. Hopefully, this book will help close the gap between the two and, in so doing, contribute toward invalidating Feynman's famous quotation printed in the opening of this book: "Nobody understands quantum mechanics."

The realm
of quanta

THE SUM TOTAL of scientific knowledge amassed since the beginning of our present scientific age is truly impressive. From the nature of the force that binds quarks together to form protons and neutrons, to the hidden codes of genetic information spelled out on a string of bases on a DNA molecule, to the discovery of the formation of a galaxy in a far corner of the universe, we have managed to accumulate a great body of knowledge about the physical world that surrounds us. Even more impressive, perhaps, than the amount of knowledge is the relatively short time span in which it was gathered. Considering that it was 2 million years ago that the earliest humans began hanging around the surface of the earth and that recorded history goes back more than 5,000 years, the history of science—from the days of Galileo and Newton, through Einstein and to this date—is a humblingly short one of only 400 years.

The year 1900 serves as the dividing line between the modern physics of the twentieth century and the earlier era of classical physics, a period spanning the years 1600 to 1900. The work of Galileo Galilei, which established the first analytical approach to the motion of physical objects, and the monumental laws of motion laid down by Isaac Newton might sound like ancient tales today, but they go back a scant 350 years: Galileo's work in 1638 and Newton's in 1687. *Newtonian physics*, or classical mechanics as it is called in the trade, is one of the two great theories that constitute classical physics, the other being the physics of electricity and magnetism. Whereas classical mechanics is largely the work of a single man, Newton, electromagnetism was developed by about a dozen physicists over a period of about 100 years—from the 1780s to the 1880s—before arriving at its final form.

Modern physics also consists of two great theories: relativity and quantum mechanics. They have been spectacularly successful in providing us with insight into the workings of a wide range of phenomena from atoms, molecules, and solids to the world of atomic nuclei, elementary particles, and the origin of the universe. Modern physics traces its genesis to 1900. On December 14 of that year, Max Planck presented his epoch-making paper at a meeting of the German Physical Society, proposing for the first time a concept he called the quantum of light. (Strictly speaking, the year 1900 is not a part of the twentieth century, since it was the last year of the nineteenth century, but this is a minor point we conveniently choose to overlook when we refer to modern physics as the physics of the twentieth century!)

An amusing historical parallel exists between classical and modern physics. As with Newton's work on classical mechanics, one branch of modern physics, the theory of relativity, was the work of a single man, Albert Einstein. Quantum mechanics, however, took 30 years to emerge—from 1900 to 1930—having gradually evolved through the contributions of a dozen great physicists. This is similar to the development of classical electromagnetism. The birth of quantum mechanics was not an easy one, and some claim it still carries the strain of a difficult delivery. Many of its conceptual and philosophical implications continue to be debated to this day.

What distinguishes modern physics from that of the classical period is more than just a point in time. The realization that space, time, and speed are all intertwined according to the theory of Einstein was as revolutionary as the complete denial by quantum mechanics of the age-old concept of point particles of matter and the deterministic description of nature. The conceptual basis of modern physics runs counter to our experience, common sense, and intuition, all of which are based on the Newtonian view of the physical world surrounding us. It comes as a shock to learn that the Newtonian concept of an absolute space and time— something we have come to accept as a self-evident truth—is only an approximate picture of space and time, which are relative to each other. The breakdown of the old laws of motion, moreover, when applied to the ever-smaller dimensions of atoms and the resulting appearance of some totally strange rules that replace them, manage to bring out some of our fiercest psychological resistance. But such is the reality of the twentieth-century physics that we call modern physics.

Where and on what scale does classical physics end and modern physics begin? In what range of speeds do we leave behind the world of Newton and enter that of Einstein? How small is small when we speak of the microcosm that is the realm of quanta? Where is the demarcation line between the classical and quantum mechanics? Before we begin our journey into the weird realm of quantum world, let us first ponder these questions.

Mechanics, the slow and the fast of it

In everyday usage the word mechanics is a loosely defined metaphor, as in the mechanics of a foreign policy—the technique of implementation—or the mechanics of a proposal review—the procedures involved. In physics, though, mechanics carries a sharply defined meaning: It refers to a branch of physics, the oldest of them all, that deals with the action of forces on material objects and the nature of their subsequent motions. Things move in this world: the swing of a golfer and the gracefully arching path of flight of a ball . . . the drama of a lunar module carefully inching toward the Apollo command module high above the lunar landscape. Things move because forces make them move, and they move in accordance with the analytical laws of mechanics.

Building on the basic concepts of mass, space, and time, Isaac Newton in 1687 laid down his celebrated laws of mechanics, a set of clearly defined rules for speed, acceleration, momentum (mass multiplied by speed), force, and energy. As the Voyager space probe soars past the far planets and sails out of the solar system, it is awe-inspiring to know that its every move, its every adjustment to speed and course, are executed in complete accord with the laws of motion laid down 300 years ago. Until the year 1905, classical mechanics, as the mechanics of Newton is called, reigned supreme over all science. Its power to explain the dynamic behavior of nearly everything under the sun was so complete and universal that, combined with the equally impressive success of electromagnetism at that time, many scientists toward the end of the last century considered the subject of physics to be a completely accomplished matter. All that remained was merely to fill in the details.

Well, it was not to be. The reign of classical mechanics came to an abrupt end in 1905. In that year Albert Einstein put forth his remarkable discovery that space and time are not separate and independent of each other, but interdependent entities. On that basis he derived a rule for speeds that explained, once and for all, the absolute constancy (in vacuum) of the speed of light—it measures the same value whether the source of light is approaching us or receding from us. In order to accommodate the absolute nature of the speed of light, Einstein overhauled our concepts of space and time! And upon this new framework he built, single-handedly, his mechanics—the relativistic mechanics—with its own set of rules for speed, acceleration, momentum, force, and energy.

Einstein's theory was indeed a grand revolutionary development, shaking our concepts of space and time at their very foundation. But contrary to what the word ''revolutionary'' might imply, the new mechanics did not render the old mechanics invalid. Instead, Newtonian mechanics simply were overtaken and incorporated into Einsteinian mechanics.

After formulating the new mechanics, the rules of which were valid at all speeds of motion—from zero up to the ultimate speed, that of light—Einstein showed that the old rules of Newtonian mechanics corresponded to approximate expressions. Their validity was restricted to a range of speeds that is extremely slow compared to that of light—the range of speeds that is ''normal'' by our standards. What we came to take for granted as a set of absolute truths turned out instead to be only approximations of the whole truth.

This might raise a logical question in your mind: If the old is completely subsumed by the new, as is the case, then why do we still keep—and quite often use—the old mechanics? Why not simply do away with the Newtonian stuff altogether? The answer is simple and practical: We rarely have an opportunity to move ourselves, or observe things that move, with a speed that is even a fraction of the speed of light. For almost all the motions of physical objects that we deal with in our own experiences, the speeds involved are so incomparably slow that the old mechanics is more than adequate.

Let me put this point in a more quantitative perspective. The speed of light, in a vacuum, clocks in at 3×10^8 meters per second, that is, 300 million meters in one second. In 1972 a team of scientists measured the frequency and wavelength of a laser light to an unprecedented accuracy, better than one part in 300 million, and pegged the speed of light to its current value of $299,792.4562 \pm 0.0011$ kilometers per second. That translates to about 186,000 miles per second or, equivalently, 668 million miles per hour. Some speed! A mere 1 percent of the speed of light is still a mind-boggling 6.7 million miles per hour. Let's throw some numbers against this sense-numbing speed. The speed of sound in air, about 330 meters per second (the speed referred to as Mach 1) checks in at about one-millionth of the speed of light. The muzzle speed of a rifle bullet, Mach 2, clocks in at two-millionths. The escape velocity of the earth, or the minimum speed an object needs to break away from the earth's gravitational pull, is about 25,000 miles per hour, or seven miles per second, which is still a meager thirty-millionths of the speed of light.

Table 1-1 lists the so-called relativistic effects at some representative speeds. The relativistic effect refers to the degree of deviation from the calculations made on the basis of the classical mechanics, that is, the difference in results between the old and new mechanics expressed as a fraction of that obtained by the old. It demonstrates the remarkable aspect of Einstein's theory: The relativistic corrections are almost entirely negligible up to 0.01% of the speed of light, but begin to rise sharply at the mid-

Table 1-1 *The rising relativistic effects.*

At these speeds, expressed in percentage of the speed of light,	the relativistic effects, the errors incurred in classical mechanics, are
0.01%	0.0000005%
0.1%	0.00005%
1%	0.005%
10%	0.5%
50%	15%
80%	67%
87%	100%
90%	230%
99.5%	1,000%

point, reaching an error of 100% at 87% of the speed of light and shooting practically straight up beyond that speed.

So where can we draw a reasonable line to separate the Newtonian and the Einsteinian worlds? Strictly speaking, it depends on the degree of inaccuracy you are willing to tolerate. For practical purposes it is safe to draw the line, as a general guide, at 0.01% of the speed of light, which corresponds to the average speed with which the earth orbits the sun. So it is with the two schools of mechanics, the classical and relativistic, the old and new, or the slow and fast (FIG. 1-1), even though there is, strictly speaking, only one that encompasses all: Einsteinian mechanics.

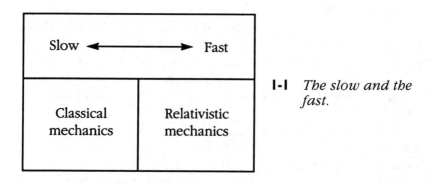

1-1 *The slow and the fast.*

Many strange consequences of the new mechanics that we often hear about, such as the slowing down of the rate of time for fast-moving clocks or the shortening of the length, measured along the direction of its motion, of a yardstick that is flying by us at fast speed, are all true and do occur. These relativistic effects, the so-called time dilation and length contraction, are everyday events in the world of elementary particle physics and high-energy particle accelerators. But as far as our own experiences are concerned, they remain as strange as they were when first introduced 90 years ago.

Mechanics, the large and the small of it

In addition to the dimension of speed, another dimension in which Newtonian mechanics reaches the limit of its validity is the dimension of size. As scientists kept uncovering layer after layer of the progressively smaller worlds of atoms and their innards— first the atoms, then to their constituents, and further down to the world of nuclei, it became apparent early in this century that the rules of classical mechanics broke down almost completely in this realm of the small. In the hierarchy of the sizes of things, there had to be a point beyond which—that is, smaller than which— totally new thinking was called for. A set of completely new concepts had to be thought out, new principles laid down, new rules discovered, and new laws formulated to explain a set of totally new facts of nature in the small. Quantum mechanics is all that. It takes up the task where the Newtonian stuff comes to a grinding halt. It is related to classical mechanics along the dimension of size, as indicated in FIG. 1-2, much the same way that relativistic mechanics is related along the dimension of speed.

1-2 *The large and the small.*

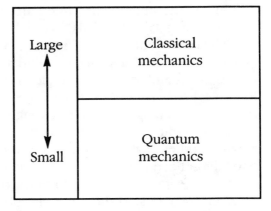

Quantum mechanics is as revolutionary as relativisitic mechanics, only more so. The revelation that space and time are intertwined is a strange notion at first encounter, but there is nothing bizarre or paradoxical about it. Quantum mechanics, on the other hand, is constructed on a new slate of foundations that are not just strange, but so bizarre and perplexing as to be almost totally incomprehensible. It is crazier than the theory of relativity. If you think about it, it serves as the vehicle of understanding and communication between us and the goings-on inside the physical world incredibly small beyond our imagination.

So how small is small and how big is not so small? Is there a point in the dimension of size that separates the domain of the large from that of the small as we have indicated in FIG. 1-2? Let's first get acquainted with some standard units often used to describe objects that populate this world of the small. Table 1-2

Table 1-2 *The powers of ten.*

Prefix	Power	Decimal	Name
tera	10^{12}	1,000,000,000,000	trillion
giga	10^{9}	1,000,000,000	billion
mega	10^{6}	1,000,000	million
kilo	10^{3}	1000	thousand
milli	10^{-3}	0.001	one thousandth
micro	10^{-6}	0.000001	one millionth
nano	10^{-9}	0.000000001	one billionth
pico	10^{-12}	0.000000000001	one trillionth

lists some of the standard notations for the powers of 10. The four prefixes in the middle are familiar from everyday usage: mega, kilo, milli, and micro. They often are used in nonmathematical contexts, such as megatrends, microcontrol, kilokill, and so on.

The other four prefixes are less familiar. A gigabit computer memory chip is an integrated circuit that stores one billion bits of data as the 1s and 0s of binary logic. A laser-optical memory disc, the so-called CD-ROM (compact disc, read-only memory), is already in this range. It won't be long before the capacity of the currently popular magnetic hard discs also reaches the gigabit range. Another relatively new "techno-name," gigaflops, that is, one billion floating point operations per second, is used often to measure—or, more precisely, to boast of—the processing speed of supercomputers. At the other end of the scale, a short slice of time known as one nanosecond, or one billionth of a second, is enough time for a pulse of light to cover a distance of one foot. Some of today's most advanced switching devices can turn electrical pulses on and off in a matter of a few picoseconds, or trillionths of a second!

Power-of-ten prefixes are also used outside science and technology. The biggest leveraged buyout in history, the $26 billion buyout of the RJR Nabisco Company, was a cool 26 gigabuck deal. In the same vein, Uncle Sam's accumulated deficit of some four trillion dollars is in the terabuck range.

In addition to these standardized notations are three specially defined units for length, each characteristic of a particular domain of science: a micron, an angstrom, and a fermi. Figure 1-3 lists some of these units and gives an indication as to what can safely—and only approximately—be called the dividing line between the "large" and "small."

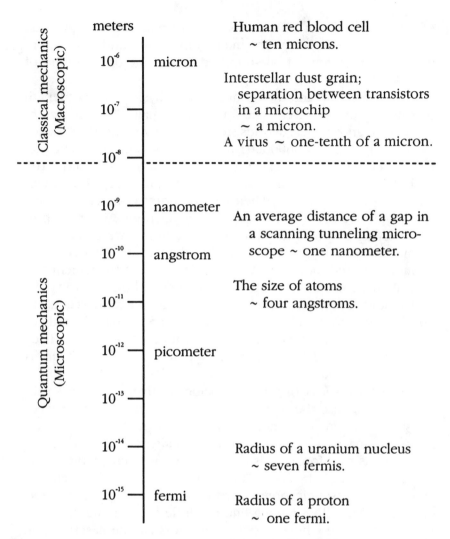

1-3 *The large and the small worlds.*

A micron is another name for one millionth of a meter, which is also called one micrometer. A slight problem with this technical designation is that the name micrometer usually means a hand-held instrument for measuring very small distances, angles, and diameters. A micron stands for the characteristic size in molecular biology and precision engineering as well as in microelectronics. Interstellar dust grains, bacteria and viruses, red blood cells, and the gap separating two adjacent transistors in today's microchips are all measured in microns, typically ranging between one-tenth of a micron to about 10 microns. An unsurpassable limit, at least with today's technology, for microchip design and fabrication, lies somewhere toward one-hundredth of a micron, the limit of so-called submicron technology.

The angstrom and the fermi are two physical units of length that belong exclusively in the quantum domain. They represent the atomic and nuclear worlds, respectively. An angstrom stands for 10^{-10} meter, 0.0000000001 meter, or one-tenth of a nanometer. It is a typical length scale for atomic physics. Out of some 105 different species of atoms that exist in the universe, 81 are naturally occurring and stable. Their sizes do not vary much from one to the other, ranging from about 1 to 5 angstroms. When we proceed further down toward the center of atoms, we come to the atomic nuclei; they are much smaller than atoms by a factor of about a hundred thousand. At this scale, another convenient and typical unit is introduced: the fermi. Named in honor of the great Italian-born American physicist Enrico Fermi, the fermi is equal to 10^{-15} meter, a milli-picometer, or a micro-nanometer. The radii of atomic nuclei also do not vary much from one to another. They range from about 1 fermi for a solitary proton to about 7 fermis for a uranium nucleus.

When we speak of things as being relatively large ·or small, we frequently use the prefixes macro or micro, as in macroeconomics versus microeconomics. But the dividing line between micro and macro is a relative one and not always sharply defined. The same is true for what is macroscopic or microscopic. Literally, macroscopic objects are those large enough to be seen and handled; microscopic objects are those invisible to our naked eye. In physics, the dividing line is a little sharper. Macroscopic objects are those whose physical behaviors can be described by the rules of classical mechanics; microscopic objects usually require quantum descriptions. The dividing line is not precise, but it is safe to say that it lies in the vicinity of 10^{-8} meter. At about 10^{-9} meter a phenomenon called quantum tunneling, one of the main subjects of this book, is observed. On the other side of the fence are viruses as large as about 10^{-7} meter. So this is about where the domain of quantum mechanics begins.

So far as nature deals out to us the systematic hierarchy of layers upon inner layers of structures—molecules made up of atoms, atoms out of electrons and nuclei, atomic nuclei out of protons and neutrons, and they in turn out of quarks—quantum mechanics is the only tool with which we can probe and expand our knowledge of matter at its deepest level of constituents. Out of this discipline will have to come answers to our unending quest for the identity of the ultimate building blocks of the universe. Out of it, too, will come the cutting-edge advances of the next century's technology. Such is the preeminent role of quantum mechanics.

The founding of quantum mechanics

The creation of quantum mechanics was the result of combined efforts by scores of physicists spread over a period of 30 years, from 1900 to 1930. During this time there unfolded a string of experimental discoveries of new and strange phenomena on the one hand, and a set of bold and radically new theoretical hypotheses on the other. New data, totally inexplicable by the rules of classical physics, would give rise to some far-out theorizing. As some of these new hypotheses were confirmed by further experimental tests, physicists started looking for underlying principles and a general theoretical framework in which to formulate a set of new rules and equations, all of which culminated in a completely new physical theory we now call quantum mechanics. In view of the fundamental and profound changes that it has wrought in the way we define the realities of the physical world surrounding us, what is surprising is not that it took so long, but rather that it only took 30 years.

During this period, too, some of the most sacrosanct concepts of classical physics had to be completely overturned. The first and the foremost casualty was the sharp and mutually exclusive distinction between a wave and a particle. Several crucial experiments proved beyond doubt that, in the small world of atoms, a physical object somehow managed to possess a dual characteristic, sometimes appearing as a particle and at other times, just as convincingly, behaving as a wave. Two concepts as diametrically opposed to each other as a spread-out wave and a pinpoint particle somehow had to merge and, in the end, within the reduced dimensions of the world of atoms, became confluent. More than anything else, this mysterious duality is the defining foundation of quantum mechanics.

Another cherished concept that fell by the way was the assumption that any physical measurement could be carried out to an unlimited accuracy with complete certainty about its out-come. Nothing in classical physics placed any limit, in principle, on the degree of accuracy with which we could measure such quantities as position, speed, a moment in time, or energy. How-ever, the wave-particle duality cannot entertain such certainty and accuracy; they are replaced instead by a new set of rules that spells out the probabilistic nature and consequently uncertain outcome of measurements. These three radical departures—the concepts of duality, probability, and uncertainty—form the con-ceptual and philosophical foundation of quantum mechanics. How did they all come about? Let's take a brief look at the birth of quantum mechanics, paying special attention to the three found-ing principles.

Quantum mechanics was born in 1900, when Max Planck of Germany (1858−1947) announced his strange idea, a radically new proposal that not only contradicted all known classical phys-ics, but also stood without any logical theoretical basis. Empirical as it was, it did have one unassailable experimental support: It brought a long-standing puzzle to a satisfactory solution. For sev-eral years a serious discrepancy between an experiment and a the-ory had existed: An observed pattern of energy emitted by the glow of a heated object, plotted as a function of frequency, could not be explained by any calculations based on classical physics.

Max Planck, in studying this situation, came upon the thrill-ing discovery that a perfect explanation of the data could be made if he only allowed himself the luxury of entertaining one very far-out notion: that the energy of radiation, be it heat, microwave, infrared, light, or whatever, is not continuously valued as had been taken for granted in classical physics. Rather—and this is the shocking part—it comes in discrete multiples of some basic and indivisible unit of radiation energy. The energy of a radiation might be 1, 2, 70, 600, 4 million, or even 10 million trillion times the energy of these basic units, but it is nonetheless a countable sum of them. The quantity of energy carried by these newly hypothesized units of radiation, which Planck named the quanta of radiation, is too imperceptibly miniscule for us to have noticed in our macroscopic world. A single penny is a ridiculously insig-nificant amount of money when talking about Uncle Sam's bud-get, a few $10 billion here and a few $100 billion there, but these monies are still countable sums of just so many pennies. And so it was with the energy of radiation, coming in a packet of a countable number of quanta of radiation.

The idea of the quanta of light, the basic units of discrete energy of radiation, remained a strange but useful notion until it was picked up in 1905 by Albert Einstein (1879−1955), who applied the idea with remarkable success to another inexplicable experimental situation involving what we now know as the photoelectric phenomenon. Not only had he achieved another triumph for the quantum idea, but Einstein went one step further with it. Apparently taking the idea much more fundamentally than perhaps Planck himself, and clearly sensing its deeper significance, Einstein elevated the concept of quanta to a full-fledged status of bona fide "particles" of light. These particles, which Einstein called photons, have no mass because radiation is massless. In one bold stroke, Einstein extended the concept of particles to cases without mass. (The notion of particles without mass runs into a wall of resistance from my students every time I introduce photons.) The work of Planck and Einstein established the beginning of the particle theory of light, in which the classical wave of radiation turns out to be, when examined in the atomic scale of the microscopic world, a shower of particles.

It was fully 19 years after the photon theory of light was established that someone dared to propose the flip side of the wave-particle duality, namely, that particles should possess a wave-like property. The idea was as preposterous at the time it was proposed as it was purely speculative. In 1924 Louis de Broglie of France (1892−1987) did just that. Without a shred of experimental evidence to back it up, he proposed it anyway, in, of all places, his doctoral dissertation. He based his conclusion entirely on what he perceived to be the balance in reciprocal symmetry of nature: If a classical wave should turn out to behave as a hail of particles, e.g., photons, then something we had always taken for granted to be a classical particle, e.g., an electron, should perhaps possess a wave-like property—at least within the domain of the microscopic world. Had it not been for a favorable reaction to the idea by none other than Einstein himself, de Broglie might never have received his doctoral degree. As it happened, this matter-wave hypothesis was dramatically confirmed only 3 years later. Two physicists at Bell Laboratories in the United States, Clinton Davisson and Lester Germer, were able to measure the wavelength of an electron wave. This was the beginning of the celebrated matter wave and the true beginning of the wave theory of matter, the quantum mechanics for matter particles.

The photon theory of light put forth by Planck and Einstein, on the one hand, and the theory of matter waves by de Broglie, on the other, constitute the principle of wave-particle duality, the first principle of quantum mechanics. However, this principle raised more questions than it answered. The first and the foremost was one of interpretation: Just how was one to make sense out of this seemingly contradictory dichotomy? The founding fathers of quantum mechanics struggled with this question for more than 30 years.

The struggle to interpret this wave-particle duality split the founding fathers into two opposite camps. One camp, led by Niehls Bohr of Denmark (1885−1962) and Max Born of Germany and later England (1882−1970), advanced what was to become the probabilistic interpretation of the duality, which asserts that the wave characteristics of an object harbor a mathematical information on the probability with which the object would be found—observed, detected, or measured—at a particular position. This interpretation completely wiped out the precise accuracy and determinability assumed inherent in classical physics for the previous 300 years. As such, it was vehemently opposed by the other camp of founding fathers, which included de Broglie, Einstein, and Erwin Schrödinger. To his dying day, Einstein could not bring himself to accept such a probabilistic view of nature, a view made famous by perhaps his most-quoted comment, "God does not play dice." In the end, however, the Bohr-Born interpretation of duality gradually established itself as one of the founding pillars of quantum mechanics.

The wave-particle duality and its probabilistic interpretation led to one immediate consequence: an inherent uncertainty in the measurement of the position of a particle. Since the occupation of a particular position is determined only up to a certain probabilistic distribution—20% chance to be at position A, 35% probability to be at a position slightly to the right of position A, and so on—a set of repeated measurements on a same system yields different results. This uncertainty in measurements of a physical object, owing to its dual characteristics, therefore is an inherent aspect of its very definition. The principle of uncertainty enunciated in 1927 by Werner Heisenberg of Germany (1901−1976) is in this sense essentially the statement of this new quantum fact. The three principles—duality, probability, and uncertainty—are then the founding bedrock of quantum mechanics that contain all of its mysteries. In the next four chapters, I discuss them in detail, and, to set the stage, I review in the next chapter how different from each other the concepts of a particle and a wave are within the framework of classical physics.

2

＊

Grains are grains and waves are waves

THE WAVE-PARTICLE DUALITY of quantum mechanics, the double "personality" of things in the microworld, was not invented by physicists. It certainly was not something wished for, but was thrust upon us by a series of crucial experiments carried out early in this century. Depending on the nature of the probes, a hail of electrons behaved at times just as a bunch of particles should. But at other times electrons acted in a way that only a train of waves could. Even after generously allowing for the fact that we are dealing with physical properties on an unimaginably small scale, this duality looks so contradictory and paradoxical that it rattles the very core of our common sense and intuition. After all, we have long established a clear-cut and well-differentiated picture of what a particle and a wave should be. Before journeying into this "twilight zone" of duality, it is worthwhile to review, reconfirm, and take stock of what is meant by a particle and a wave in the classical sense. In so doing, we search for what little common ground might be shared by the two concepts, even within the framework of classical physics.

A good old classical particle

The notion of a particle is so familiar that it is difficult to find something to say about it that is not already obvious. A grain of sand, a speck of dust, a tiny bit of matter that presumably has a definite shape, preferably a spherical ball, has a definite mass and a hardness we can feel. This is the matter the universe is made of. A particle is a particle and that is that; every child knows what it is. Abstracted out of such an intuitive picture is the classical definition of a point particle, a bit of matter having a definite mass that occupies a mathematical point in space. Under the influence of a force, it traces a unique path of motion. From the beginning of science the point particle has been the basic unit of matter used to analyze the physical world surrounding us.

The foundation of Newtonian mechanics rests on three basic concepts—mass, space, and time. All other physical quantities, such as velocity, momentum, acceleration, force, and energy, are derived from them. The mass defines matter, which consists ultimately of mass points, to wit, point mass particles. A mass in turn is a measure of a property called inertia, as Newton had originally defined it, arising from the property of matter being inert. That is, it resists any change in its status of motion. In *Principia*, Newton defined mass to be the quantity of matter, where ''the quantity of matter is the measure of the same, arising from its density and bulk conjointly.''

A particle occupies one position in space at any moment in time, tracing out as it moves a unique trajectory of motion specifying its position as a function of time. The rate of change of position in time, that is, the slope of the trajectory, defines the velocity of a particle. The velocity multiplied by the mass defines its momentum. The slope of a velocity curve, that is, the rate of change of velocity in time, is what we call acceleration. Newton's second law of motion states that the net force exerted on a particle is equal to its mass multiplied by its acceleration. A force changes the status of a particle, speeding it up, slowing it down, changing the direction of its motion, and constantly changing its location. The measure of the effect of a force is expressed as changes in the energy of a particle. Whereas the position, velocity, and acceleration describe the kinetic aspects of the motion of a particle, the dynamic impact that it imparts to other particles is characterized by its momentum and energy. And all the kinetic and dynamic aspects of a particle can be determined, in classical physics, with complete predictablity—and, in principle, unlimited accuracy. This is the essence of Newtonian mechanics, which is built on the basic tenet of a point mass particle.

That a point particle occupies one definite position at one time can be viewed from a slightly different perspective. When we do this in terms of the conventional classical theory of probability, the picture that emerges might surprise you. It does seem to provide a window for a new way of looking at things that are old and familiar—a small, narrow opening, but a window nevertheless.

Suppose we have a particle, say a tiny ball at rest on a table top. In some convenient scale, let us say that the particle is located at the position +2, as shown in diagram A of FIG. 2-1. This seemingly trivial statement can be rephrased by stating that the probability of the particle being at that position is 100% while the probability of it being anywhere else is 0%. If we plot this probability distribution, we get the "curve" shown in diagram B. This horizontal flat line represents zeroes, or the zero probability for everywhere except at the position +2, where the vertical spike of height one indicates 100% probability. Viewed mathematically, that is one heck of a "curve," too discontinuous and certainly too extreme.

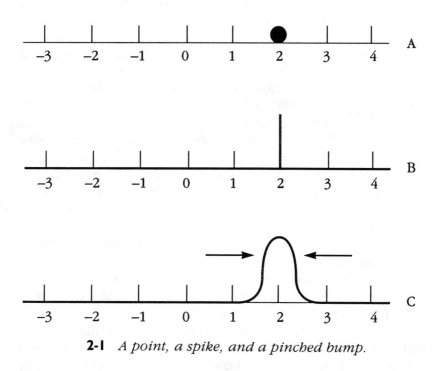

2-1   *A point, a spike, and a pinched bump.*

Had it not been for a serious difficulty in its physical interpretation, it would seem more natural to consider the sharp spike as the limiting case of a narrow bump, a continuous and smooth-varying function, as it is squeezed in from both sides, as depicted in diagram C. As a bump in the probability curve, however, such a point of view cannot be entertained in Newtonian mechanics. A particle's location cannot be pinpointed with certainty except to say that it is confined within the space between the positions 1.5 and 2.5, and the bump represents the probability distribution of locating the particle at any point in between. No such probabilistic distribution and the ensuing uncertainty are allowed within the rigor of classical physics. This simple exercise, however, does stretch our imagination and suggests there might be room, for aspects of a particle and a wave in some way to come together.

A good old classical wave

A wave is as common and familiar a notion as a particle. It is hardly something that would make us drop whatever we are doing to pause and think about it. Perhaps the first thing that comes to mind is a scene of rolling waves at a seashore—a train of surfs roaring onto a beach, one after another. Or you might think of a family of rings of ripples on a quiet pond, gently spreading out in ever-widening circles. A wave moves—spreads, travels, or propagates—in all directions with its own characteristic speed. A ripple on a pond might travel only a few feet per second while some ocean swells are known to travel far and fast. Powerful waves caused by an underwater earthquake near Alaska traveled all the way to the southern tip of South America at speeds up to 50 miles per hour. Next to water, the wave we are most familiar with is sound. The most common form is the sound wave in air, although sound also travels through liquid and solid as well.

Both water and sound waves are examples of what is called the mechanical wave, a wave that requires a material medium for its propagation. The simplest example of a mechanical wave that we can readily visualize is a bump on a piece of rope. Suppose you hold one end of a rope and the other end is fastened to a wall. A sharp up-and-down jerk at one end produces a wave pattern, a bump, that travels down the length of the rope, is reflected at the wall, and comes moving back. A few waves, that is, the snapshots of a few traveling waves, are shown in FIG. 2-2. Among an infinite number of possible patterns, the figure shows three examples: a simple bump, a cluster of waves, and an infinitely periodic harmonic wave that corresponds to a sine or cosine function of trigo-

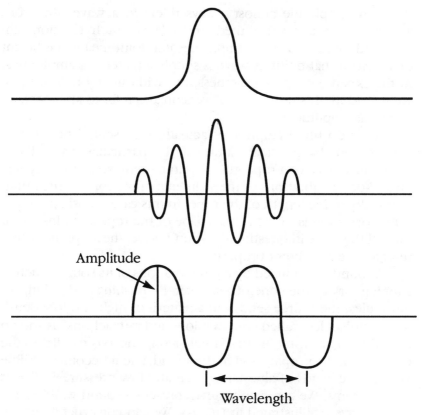

2-2 *Snapshots of a few waves.*

nometry. One basic difference between a wave and a particle is glaringly obvious: spatial extension. A particle is defined only at a point, at one time, whereas a wave knows no such bounds. A wave can be only a few angstroms long, or a few feet, or it can extend from one side of an ocean to the other, or from one end of the known expanse of the universe to the other.

The parameters we need to describe a wave differ from those used for a point particle. The distance between two adjacent crests or troughs is called the wavelength, and the number of complete oscillations in a second is the frequency of the wave. The vertical height, or depth, of a wave at any point is called the displacement at that point. One particular displacement, the maximum displacement, is defined as the amplitude of a wave. This definition of the displacements and amplitude is technically correct, but, perhaps due to the fact that the word displacement carries meanings other than those associated with a wave, whereas the word amplitude almost always refers to a wave, the word amplitude is usually substituted for displacement. In this nontechnical and looser usage, an amplitude at a point means the height of a wave at that point. A wave is simply a pattern of amplitudes. At the expense of some exactness, but without any loss of rigor, we will adopt this nontechnical version and refer to all heights of a wave as amplitudes.

What do these amplitudes actually represent? That answer depends on the physical nature of the particular wave. For a bump on a piece of a rope, the amplitude represents the physical up-and-down vibrations of the segments of the rope as the bump travels down the length of the rope. In this case, the shape of the wave corresponds to the actual shape of the rope at an instant of time. If the wave in question is a sound wave, the amplitudes represent quite a different property.

A sound is produced by rapid vibrations of its source, such as a tuning fork or the cone of a speaker. By pushing and pulling air molecules, the source sends out a pattern of high and low densities of molecules, called compressions and rarefactions, as shown in FIG. 2-3. The air molecules vibrate along the axis parallel to the direction of the propagation of the sound, and it becomes audible when these bands of alternating high and low pressures arrive at our eardrums. We can graphically represent a sound wave in one of two ways, as illustrated in FIG. 2-3. We can construct a density plot by dutifully dotting in all the molecules, or we can quantify the densities and plot density as a function of position. The second way gives us a smooth-varying distribution of amplitudes, a wavy wave. The amplitudes in this case, however, clearly do not represent any up-and-down motions of the medium, but instead serve as the mathematical representation of the relative densities of air molecules at each point over the region where the sound propagates. In this simple and innocuous example, in a classical mechanical situation, the amplitudes of a wave stand for a mathematical distribution of densities. This will be helpful when you confront the mystery of the wave-particle duality of the quantum world.

Diffraction of waves

Beyond the obvious differences between a particle and a wave in terms of their spatial extensions and other physical features, two classic phenomena occur only for waves, not for particles, pro-

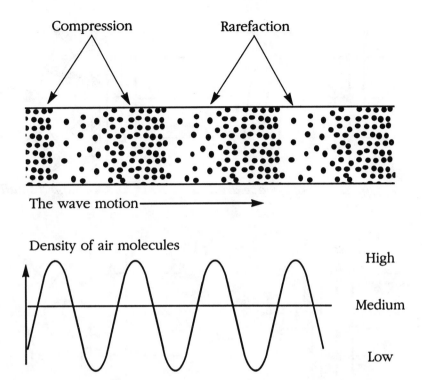

2-3 *The sound wave.*

viding an undisputable way to differentiate one from the other. The phenomena are diffraction and interference by waves. Historically, it was the observation of diffraction and interference of light that helped establish the wave theory of light. Of the two, diffraction is the more common occurrence that we experience; an intricate laboratory set-up is usually required to observe interference by visible light.

Diffraction refers to the phenomenon in which waves— water and sound waves, microwaves, and light, among others— can either bend around a sharp corner of an obstacle or spread out in all directions after going through a small opening. Two examples are illustrated in FIG. 2-4. Without necessarily realizing it, we often encounter instances of diffraction in our daily routine—getting splashed by unexpected water waves bending around a sharp corner in a swimming pool or overhearing a whisper from around an edge of a wall. Partial illumination of a shadow area is another example of diffraction. So is the fact we can still pick up a broadcast on our car radio while passing through the deepest point inside a long tunnel. The diffraction is

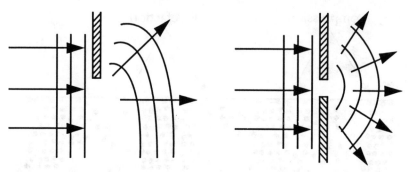

2-4 *Diffraction of a wave around a corner and through a small opening.*

NO!

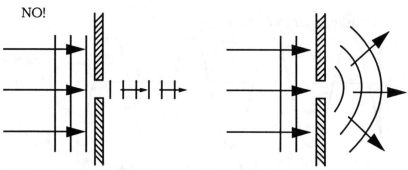

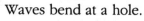

No way! Waves do not traverse straight through a hole.

Waves bend at a hole.

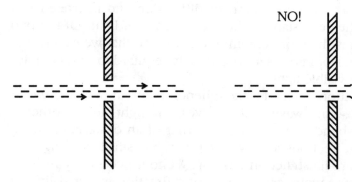

NO!

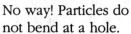

Particles traverse straight through a hole.

No way! Particles do not bend at a hole.

2-5 *What particles and waves do going through a small opening and what they do not.*

almost synonymous with a wave; if a thing diffracts, then that thing, whatever else it may be, has to be a wave. It is that simple.

Particles are different. The point mass particle, as defined in Newtonian mechanics, cannot diffract. A particle moves along its characteristic trajectory, consistent with Newton's laws of motion. In the absence of any net force to change its motion, it will remain on course on a straight path. If a beam of classical particles is shot at a small opening in a wall, those particles that get through the hole will continue on straight path. Barring collision off the sides of the opening, they will not bend their trajectories by themselves. Particles simply do not diffract, as the two upper diagrams in FIG. 2-5 show. The two lower diagrams illustrate the other side of the coin, namely the yes-and-no situation in terms of a wave incident on a small opening. As one of them shows a deliberately false situation, it would be absurd to imagine that a wave would emerge from the hole diminished in its width and continuing on a straight path! The four diagrams summarize exactly what does and does not happen when a classical particle and a classical wave come upon a small hole in a wall.

Interference, the double-or-nothing trick

All the sounds we normally hear are combinations, random or otherwise, of several different sound waves, as well as their reflections off various surfaces. Listening to music from a state-of-the-art hi-fi system, we hear the sum of sounds coming from as many as 10 speakers acting as a set of correlated sources. We can readily experience the phenomenon called the interference of waves by shifting our listening position. We hear greatly amplified sounds at some locations; at other positions we hear greatly muted sounds due to mutual cancellations. When two or more waves are put together, that is, when many waves are superimposed on top of each other over the same region of space, the amplitudes of each wave interfere with one another. At every point the resulting amplitude is the algebraic sum of all the amplitudes from each component wave. Ups reinforce ups, downs reinforce downs, and ups and downs subtract from each other. The interference of waves is universal for all forms of waves. Two separate bumps moving in opposite directions on a rope will meet each other, produce a new shape, and continue to move on in their original shapes. Drop two pebbles into a quiet pond and you can watch the dance of two expanding ripples continuously adding and subtracting from one another as they spread out in widening circles.

This interference, or the superposition of waves, is shown in its simplest form in FIG. 2-6, using what are called square waves. No real waves come in such block shapes, but the simplicity of it is ideal for illustration. At each point between the ends marked a and e, the amplitudes of the sum of two waves are the algebraic sum of the two amplitudes at that point. A slightly different arrangement of a superposition of two bumps is shown in FIG. 2-7. Place a ruler vertically over the figure and, as you move the ruler across the face of Morris the Cat, you can verify that the drawing is fairly accurate. By superimposing as many—and as many different—waves as you desire, you can generate an infinite variety of new shapes and superposition of many harmonic waves. You can create some interesting patterns for the resultant waves. In FIG. 2-8 two harmonic waves of the same amplitudes, but a 2-to-3 relative ratio in frequencies, are superimposed to create a teeth-like pattern. Three different waves are put together in FIG. 2-9 to yield a reasonable synthesis of a block wave. By adding more harmonic waves with smaller amplitudes and judiciously chosen frequencies to the three shown in the figure, it is possible to come very close to the block-like square pattern.

Things become interesting when we consider the superposition of two identical waves, exact clones of each other. Under some exacting conditions they play a "double-or-nothing" trick called constructive and destructive interferences. One simple arrangement is shown in FIG. 2-10. Suppose a wave source, either a sound wave or visible light, at location A sends out the wave moving to the right. Suppose further that another identical source at position B also sends out to the right an identical wave with the same amplitude, frequency, and wavelength. The diagrams in FIG. 2-10 are almost self-explanatory. In the first diagram the source B is located one full wavelength to the left, or to the rear, of the source A. For all points to the right of point A the two identical waves are in a perfect unison with each other. Consequently, the amplitudes of the two identical waves will reinforce each other at every point and double the amplitude all the way down to the

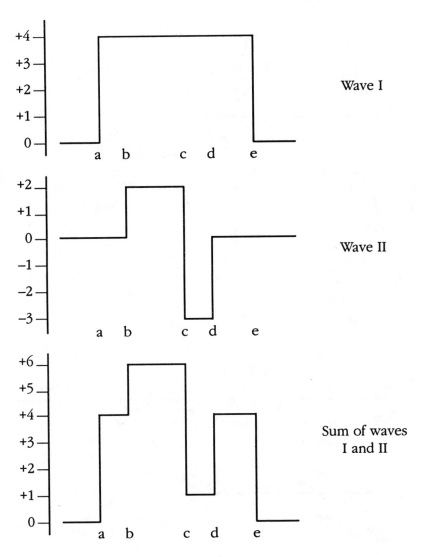

	Wave I	Wave II	Sum
To the left of point a	0	0	0
Between points a and b	4	0	4
Between points b and c	4	2	6
Between points c and d	4	−3	1
Between points d and e	4	0	4
To the right of point e	0	0	0

2-6 *Superposition of two bumps.*

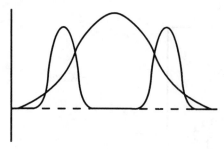

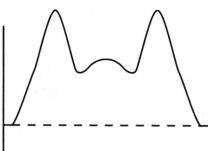

2-7 *Another superposition, Morris the Cat.*

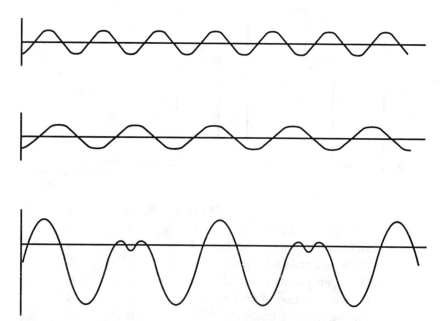

2-8 *Superposition of two harmonic waves with a 2-to-3 ratio in frequencies resulting in a "teethy" pattern.*

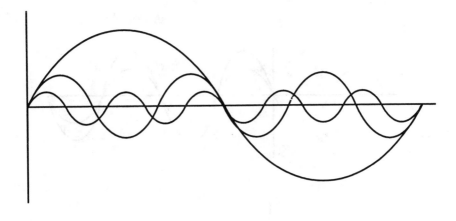

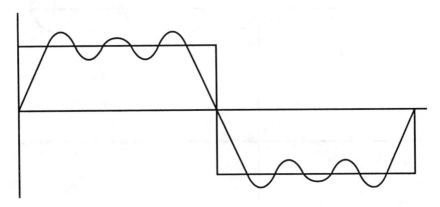

2-9 *Superposition of three harmonic waves synthesizing a "square" wave pattern.*

right past point A. Such a doubling of waves is called a constructive interference.

You don't need me to tell you how the nothing part of the double-or-nothing trick works when the two waves are arranged as shown in the second diagram of the figure. The source B is exactly half a wavelength behind the first wave. As a result, for all the points to the right of position A, the two waves are out of step with each other. Amplitudes from the two waves exactly nullify each other all the way down to the right of point A. This is called the destructive interference. The two waves have completely killed each other off; no wave is left.

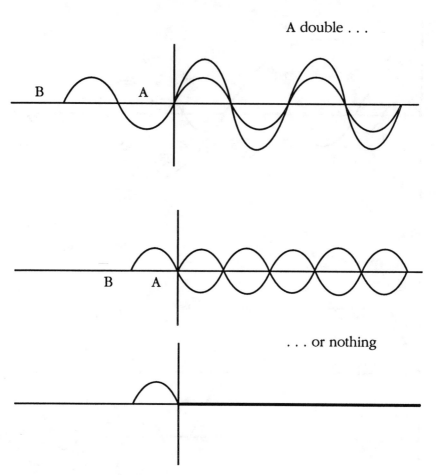

2-10 *A double-or-nothing trick.*

These types of interference, constructive and destructive, are perhaps the most dramatic of all the superpositions of waves. They provide exacting experiments in which the wave properties are accurately measured. The two classic wave phenomena, diffraction and interference, helped verify the wave theory of light 200 years ago. In the early part of the twentieth century, the same two phenomena helped usher in the wave theory of matter, or quantum mechanics, as the next three chapters discuss.

3

The shower
of light drops

THE SET OF PHYSICAL EVIDENCE that brought to light the
mysterious behavior of the wave-particle duality divides, quite
naturally, into two groups: evidence that demonstrates the particle-like behavior of what was clearly considered a wave and evidence that shows the "waviness" of what was otherwise a bunch
of particles. Almost without exception, discussion of the difficult
concept of duality begins with the first group, the particle-like
properties of a wave, for two good reasons. First, that is the way
things developed historically, and second, and more important, it
is still much easier to understand that way. After all these years,
the waviness of a particle is still a more difficult subject. After
nearly a century of quantum mechanics, particles are easier to
deal with conceptually.

The particle-like aspect of a classical wave is a quantum
effect; that is, the granular behavior of a wave shows up only
within the reduced dimensions of the quantum world. As I discussed in chapter 1, the world of quanta sets in when the scale of
things becomes smaller than the characteristic threshold. For a
wave, the quantity that characterizes the scale is its wavelength.
Just as the smallest unit etched on a ruler defines the smallest
length that the ruler can measure, the wavelength of a wave represents the smallest scale of things that the wave can gauge. The
quantum granularity of a classical wave can begin to be observed,
therefore, only for those waves whose wavelengths are short

enough to be comparable to the dimensions of the quantum world.

Obviously, we cannot expect to observe any particle-like properties of a water or a sound wave, much less a bump on a rope. The wavelengths of the mechanical waves are too large, too classical. It is only with visible light and beyond—ultraviolet, soft X-rays, and so on—that we observe the quantum granularity. This situation is symbolically depicted in FIG. 3-1; as its wavelength becomes shorter and shorter, the wave suddenly appears to turn into a hail of particles. This chapter focuses on what is called the "particle theory of the wave of light" and discusses the electromagnetic wave and its quantum particles, the photons.

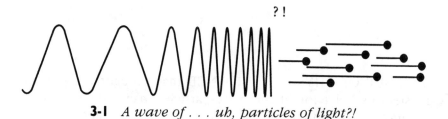

3-1 *A wave of . . . uh, particles of light?!*

Maxwell's triumph

The natural phenomena of electricity and magnetism have been observed since the earliest civilizations. The word *electricity* owes its origin to the Greek word, *elektron*, for amber, a brownish-yellow fossil resin often used to make pipe stems and jewelry. When rubbed with a piece of cloth, amber attracts dust and small bits of paper—the property we know as static electricity. Magnetism traces its history to a region of Asia Minor called Magnesia, now a part of western Turkey, where rocks were discovered that exerted a strange pull of forces to each other. The quantitative and analytical study of what is now called the classical theory of electromagnetism is, however, of relatively recent vintage; it was developed in a period of about 100 years, running from the 1780s to the late 1880s.

This investigation began in 1785 with the precise determination by Charles Coulomb of the nature of the force acting between electric charges. Called Coulomb's law, it is the electric counterpart to the law of gravitational force discovered by Newton a century earlier. The pace of research quickened when Allessandro Volta invented the first working battery in 1800. The battery enabled people to devise and carry out all sorts of tabletop experiments, some successful and others not, some scientific and others not quite so scientific. One of these experiments, using batteries and wires, led to an accidental discovery with a far-reaching consequence. In 1820 Hans Christian Oersted noticed the curious movement of a magnetic compass needle when a nearby wire was connected to a battery. The battery was a source of electric force, and as far as then known there was no reason even to suspect that electricity and magnetism were remotely related. But the two forces had to be somehow related! Otherwise, why would a magnetic needle be affected by a nearby current-carrying wire? This was the first inkling that these two aspects of nature were closely linked.

The unification of electricity and magnetism, that is, the realization that the two are opposite sides of the same coin, occurred in two steps. They happened about 30 years apart, largely at the hands of two men, Michael Faraday (1791 – 1867) and James Clerk Maxwell (1831 – 79). In 1831 Faraday, and, independently, Joseph Henry (1797 – 1878), uncovered a remarkable fact: An electric force was produced whenever the strength of a nearby magnetic force changed. The electric force came not from the charged end of an amber rod, nor from a chemical battery, but instead from increasing or decreasing the strength of a magnetic force! With a loop of wire and a strong desktop magnet, you can generate an electric force—thereby inducing a current in the wire—by moving either the loop or the magnet, or both. This, as we know today, is the principle behind such industrial workhorses as generators, alternators, motors, and transformers, but at the time, it harbored a deep theoretical implication—the oneness of electricity and magnetism.

Then in 1864, came a sweeping mathematical treatment in which James Clerk Maxwell made three great theoretical achievements: a grand synthesis of the laws of electricity and magnetism into a set of unifying equations—called Maxwell's equations—that defined classical electromagnetism as we know it today; the theoretical prediction of the existence of the electromagnetic wave, and the totally unexpected result that visible light was part of this newly theorized wave. It was a magnificent sweep, one of the greatest achievements of all time.

Building largely on the previous work of Faraday, Maxwell reasoned that the converse of Faraday's law must also be true: If a change in the strength of a nearby magnetic force produces an electric force, then a magnetic force should be produced by a change in the strength of a nearby electric force. This guess, when combined with existing laws then known, not only led to the complete synthesis of the theory, but also gave rise to an unmistakable equation that showed that a wave of undulating electric and magnetic forces existed—a new, hitherto unknown form of wave. This was not a wave on the surface of water, nor a sound wave through air molecules, nor a bump on a rope, but an undulating fluctuation of an electromagnetic force that propagates through empty space. It needed no material medium for its existence.

These two achievements, the grand synthesis of classical electromagnetism and the prediction of a totally new form of wave, would have been more than enough to have ranked Maxwell as one of the great physicists of all time. But he had yet another great discovery to make. When he calculated the speed of propagation of this newly hypothesized electromagnetic wave, Maxwell found, much to his own astonishment, that it was identical with the known value for the speed of light! Based on this remarkable coincidence, Maxwell proclaimed, without much hesitation, that visible light had to be part of this new and not-yet confirmed form of wave. In 1887, eight years after Maxwell's death, Heinrich Hertz confirmed the existence of such an electromagnetic wave by generating it in one corner of his small laboratory and detecting it in another. And so it came to pass that mankind learned for the first time what light was all about.

The electromagnetic wave

Visible light corresponds to only a small portion of what turned out to be a wide and various spectrum of the electromagnetic wave. The full range contains, in addition to light, radio and television waves, microwaves, infrared and ultraviolet radiations, X-rays, and potent gamma rays. Every portion of this spectrum is the source of immeasurable benefits to us, benefits that are not only indispensible to our well-being, but vital to our existence. As familiar as electromagnetic waves might be, however, their physical nature, that is, their actual "waviness," is not readily recognizable. We perceive light by its colors as well as its brightness, but we cannot make out its wavy pattern. What forms a wave in water is obvious. What is waving in a sound wave is perhaps a little less obvious, but it is still easily recognizable as variations in the densities of air molecules, as shown in FIG. 2-3. When it comes to the electromagnetic wave, however, what exactly is waving is not at all obvious.

First, what generates an electromagnetic wave is the electric charge, the same charge that gives rise to electric and magnetic forces. More precisely, while a magnetic force is generated by electric charges in motion—an electric current—and the electric force always acts between charges, in motion or otherwise, it is a sudden acceleration, deceleration, or a rapid oscillation of electric charges that creates a wave of electromagnetic forces. These forces spread out, or propagate, through space in much the same way that water waves spread out in expanding circles when a pebble is dropped into a pond. Rapidly oscillating currents in a broadcast antenna of a transmission tower beam out emergency police and fire calls, radio and TV programs, and microwave transmissions. Electromagnetic waves are also emitted, as well as absorbed, by atoms and atomic nuclei, especially in the bands of visible and ultraviolet light, X-rays, and gamma rays. Such atomic and nuclear emissions also fall under the general category of accelerating or oscillating charges because the electrons, going around in circles inside an atom, are executing, an oscillatory motion. The frequencies of visible light being in range of 10^{14} oscillations per second implies, among other things, that electrons inside an atom emitting a light are going around in circles at the rate of 100 trillion times per second!

What spreads out from an oscillating source is essentially a harmonic wave pattern of fluctuating strengths of the electric force plus an accompanying wave of magnetic force. Two such waves are illustrated in FIG. 3-2. As wave 1 travels to the right, what travels is the variation in the strengths, as well as in the directions, of the electric force. The upward maximum is at point b, and the downward maximum is at point d. The electric force is always pointed, either up or down, perpendicular to the direction of propagation. The directions of the electric forces in wave 2 illustrate this point.

At this point we begin to see the wonders of the discoveries made by Faraday and Maxwell. As the electric force fluctuates, the

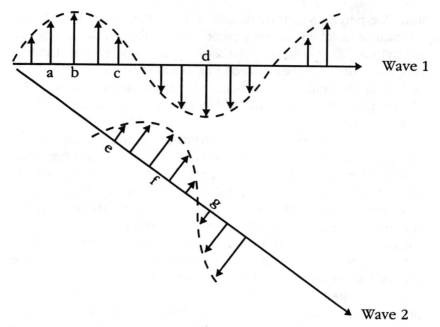

3-2 *Waves spreading out from a source. The waving strengths and directions of the electric force at points along the line of propagation are illustrated.*

changes generate, along the way, a magnetic force that is also fluctuating. The changing magnetic force, in turn, regenerates the electric force! So they go, generating and regenerating themselves, propagating out in all directions in a form of a wavy dance that we call the electromagnetic wave, from a nearby transmission tower to our homes, or from one end of the universe to the other. The electromagnetic wave needs no material medium in which to wave—no water, no air, nothing. It waves itself and is completely self-perpetuating.

The spectrum of the electromagnetic waves schematically shown in FIG. 3-3 covers the full range from a very long wave to some of the most powerful gamma rays known. In terms of either the frequencies or wavelengths, the range spans some 22 orders of magnitude, from 10^2 to 10^{24}. Any electromagnetic wave has only one unique speed with which it propagates: the speed of light, which in vacuum is 186,000 miles per second or 300 million meters (3×10^8 meters) per second. Since the speed of a wave is equal to its wavelength multiplied by its frequency, the longer the wavelength of a wave, the lower its frequency. Conversely, the

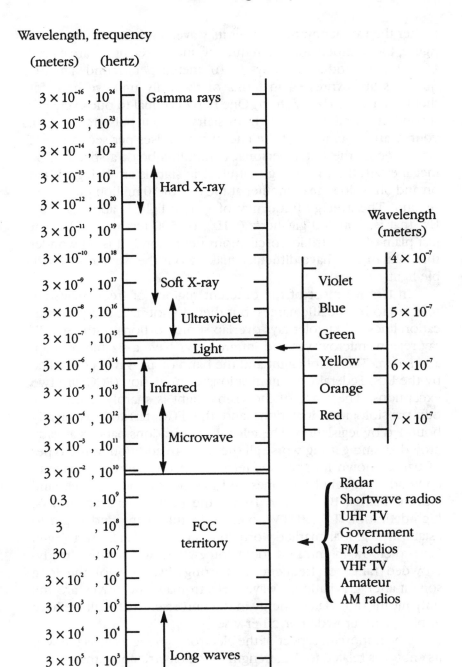

3-3 *The electromagnetic spectrum.*

higher the frequency, the shorter its wavelength. As shown in the figure, the numbers for the frequency and wavelengths are paired so that their product is always 3×10^8 meters per second. The frequencies are expressed in terms of the only unit we have for them—hertz, or Hz for short. One hertz is equal to one cycle per second. After climbing a flight of stairs, if your pulse rate is 120, your heart is pumping at the rate of two pulses per second, or 2 Hz. Some of the faster personal computers boast a speed of 30 megahertz (MHz), meaning that electric signals are being turned on and off inside the computer at the rate of 30 million pulses per second. The average frequency of visible light is about 5×10^{14} Hz, that is, 500,000 gigahertz (GHz), or 500 terahertz (THz), or just plain old 500 trillion oscillations per second. It is no wonder that human eyes have difficulty making out the waviness of visible light.

In FIG. 3-3 most of the different regions of frequencies are referred to by specific names, but they do not have sharp demarcation lines. Ultraviolet rays overlap some portions of the soft X-ray region, microwaves and infrared rays have a sizable overlap, and so on. The visible light and the band of air waves controlled by the U.S. Federal Communications Commission (FCC) are two exceptions: The range of the visible light is sharply defined by our physiological limitations and the FCC territory is clearly bounded by legislation. The telecommunications band is already crowded, and getting worse all the time. In addition to the types of bands shown for the FCC territory in the figure, we have to contend with cellular phones and marine, aviation, police, and citizens bands! By the late 1990s, the next generation of TVs, high-definition TV (HDTV), will be in full swing and begin to establish a global market worth billions of dollars. HDTV will have to elbow in and find room somewhere within this already-crowded range of frequencies. Barring discovery of the long sought-after gravitational waves, electromagnetic waves are the only means of radiated energy in the universe. That is all we have to play with, crowded or otherwise.

One important aspect of the electromagnetic wave is the way its energy is related to the strength of the electric force that makes up the wave pattern. It is simple and straightforward: It takes more energy to generate a larger amplitude and the same energy for the same amplitudes, whether up or down. An elementary calculation yields the expected result, namely, that the energy at every point along the direction of propagation is proportional to the square of the height, or the depth, of the wave at that point. This is shown in FIG. 3-4, where the vertical scales for the energy and the wave are not identical, but only proportional to each other. This simple aspect, seemingly quite innocent, will be a crucial ingredient later in the discussion of the probabilistic interpretation of the wave-particle duality.

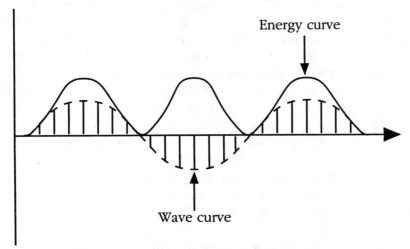

3-4 *The energy is proportional to the square of the amplitudes.*

Photons, the pellets of light

Now we come to the crux of the matter—the discovery of the "graininess" of light and the subsequent evolution of the concept of photons—the pellets, granules, in fact, the particles of light. This concept is what started the whole ball rolling toward the birth of quantum mechanics. As you can imagine, the idea of a light wave turning out to be a hail of shots did not come easy. It took 16 years, from 1900 to 1916, for the idea to solidify.

Toward the end of the nineteenth century, two new experiments began to show signs that perhaps not all was well with the grandiose theory of the electromagnetic wave. When applied to the goings-on inside the microscopic world of the small, the experiments showed that the theory might be seriously flawed. In studies of the patterns of energy radiating out of hotly glowing objects, the observed correlation between the amounts of radiated energy and the temperature of the source defied all attempts to explain it in terms of classical theory. The disagreement between the theory and the data was serious, and it persisted while many scientists expended much effort to resolve this crisis.

One such scientist was the German physicist Max Planck, who in 1900 arrived at the epoch-making conclusion: The only way to resolve the disagreement was to conclude that the energy of the electromagnetic wave, instead of having the usual unrestricted and continuous value, came in discrete amounts corresponding to some number of indivisible basic units of energy—1, 2, 500, 10,000 or even a few billion of them—but a countable sum nevertheless. It was a drastic and truly revolutionary conclusion, a direct contradiction to the classical theory, but it provided the only means of resolving the disagreement between the theory and data. It had to be accepted. As Planck reported his analysis in December 1900, he referred to this discrete unit of energy as the quantum of light.

The idea of light quanta was so radically different, however, that Planck himself considered it at first to be a transient explanation—a clever mathematical trick that happened to yield the right answer. Five years later, in 1905, the quantum hypothesis was picked up by Albert Einstein, who brilliantly applied it to another nagging problem that defied the conventional explanation, the so-called photoelectric effect. In this phenomenon, which is the everyday mechanism for opening automatic doors at hospitals and airports, a bunch of electrons are liberated out of the surface of certain types of metals when light shines on it. The energy of light "boils" out, so to speak, electrons from the metal surface, and here too the energy of light did not follow the rules of the classical wave theory.

Not only did Einstein explain the photoelectric effect, but he went one step further: He perceived these quanta of energy to be full-fledged particles in their own right, specks of light he named photons. According to Einstein, the wave of light was actually a stream of photons interacting with individual electrons on a one-on-one basis, similar to moving billiard balls. In 1916 the great American experimenter Robert Millikan carried out a precise and definitive confirmation of Einstein's theory of photons. Soon thereafter followed, in rapid succession, three Nobel prizes in physics for contributions establishing the photon theory of light: to Max Planck in 1918, Einstein in 1921, and Millikan in 1923.

In the process of elevating Planck's idea of the quanta of light to the status of *bona fide* particles of light, Einstein extended, in one bold stroke, the concept of a particle much beyond what was originally intended; he generalized it to include those cases having no mass at all. Electromagnetic radiation does carry its own form of energy and momentum, the radiation energy and the radiation momentum, but it certainly does not have any mass. As basic particles of light, photons are by definition totally massless. What Einstein did, in effect, was liberate the concept of a particle from that of mass and redefine it by its dynamic attributes as a localized lump of discrete energy and momentum. In this broader definition, photons and electrons are just as good particles as each other. Since this original definition of a zero-mass particle in 1905, a few more species have joined photons as members of a handful of zero-mass particles to date, including such particles as neutrinos and gravitons. Gravitons, if they exist, would be the gravitational counterpart of what photons are to the electromagnetic wave.

The energy of photons, however, is too miniscule to register at our human-sized macroscopic scale, and we can no more directly perceive the grainy nature of light than we can discern its waviness; there are simply too many photons in sunlight or other common light sources. Sunlight reaching the surface of the earth, for example, delivers on the average about one kilowatt of power per square meter. For one square centimeter, that is, about 3/8 inch by 3/8 inch, this translates to a steady stream of about 3×10^{17} photons striking the area each second, or 300 million billion photons. The next time you are sunning yourself at the beach, you might contemplate this onslaught of photons on your body. At the other extreme, researchers have confirmed that the human eye can see a flash of light as faint as only half dozen photons, proving the extreme sensitivity of our own optical nerve system.

When Planck originally proposed the idea of quanta of light, he made a clean break from the classical electromagnetic theory in more ways than one. Not only had he introduced the idea of discreteness in radiation energy, but he also discovered a fundamental quantitative relationship that determines the amount of energy of a photon. This relationship, known as Planck's formula or sometimes the Planck-Einstein formula, states that the energy of a photon is proportional to the frequency of its radiation. This means that the energy of photons is different for radiations with different frequencies, the photon of a radiation of higher frequency being endowed with more energy than the photon of a radiation with lower frequencies. The total energy of a radiation is hence the total countable sum of the energies of the total number of its own photons. As a relationship that connects the energy of a particle to the frequency of the wave that it is the particle of, it is a strange, hybrid relationship, that rests on the dichotomy of a wave and a particle.

Of the entire electromagnetic spectrum shown in FIG. 3-3, the portions of radiation that we are concerned with begin at the higher end of the infrared and cover everything above it, that is, visible light, ultraviolet, X-rays, and gamma rays—the range of radiation whose wavelengths are small enough to matter in the quantum world. It covers a wide range of frequencies, from 10^{14} hertz up to 10^{24} hertz. In principle, the spectrum has no upper

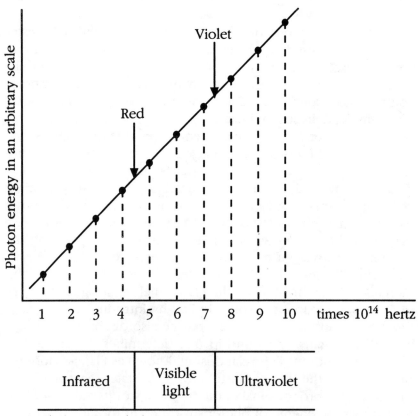

3-5   *Photon energies as a function of frequencies.*

limit, but so far we have not directly detected photons of frequencies much higher than 10^{24} hertz. A photon of red light has energy that is less than that of a blue photon. From the relatively "low" frequency of 10^{14} hertz to the "high" of 10^{24} hertz, in fact, the spread in frequencies, and hence in the energies of photons, covers a 10 billion to 1 ratio: a single photon of a gamma ray with a frequency 10^{24} hertz, which is the smallest unit of energy for that gamma ray, has an indivisible energy 10 billion times that of a single photon of an infrared radiation of frequency 10^{14} hertz. In FIG. 3-5 the portion of the spectrum centered at the visible light has been enlarged. Along the horizontal base are marked frequencies from 10^{14} to 10^{15} hertz; photon energies are plotted vertically, in an arbitrary scale, to show the relative magnitudes involved. The spread in energies for the visible light is relatively narrow, a violet photon having an energy about 1.7 times that of a red photon.

4

A breeze of
electron waves

THE DEVELOPMENT of the wave theory of a particle, the flip side of the wave-particle duality, evolved along a path in exactly reverse order from that of the photon theory of light. The particle nature of the electromagnetic radiation, as I discussed in the previous chapter, was experimentally discovered first and theorized later. It was the gross discrepancy between what was observed and what was predicted by classical physics that drove Planck to propose and Einstein to establish the particle nature of light. The wave theory of a particle, on the other hand, began with a bang of a purely theoretical hypothesis, which came to be confirmed by experiment later. Out of the blue, Louis de Broglie proclaimed that a particle ought to have wave-like properties.

The particle in question was the electron, at the time the only well-established member of what was to become the family of elementary particles. It was only three years prior to the work of Planck that electrons were discovered, in 1897, by J.J. Thomson (1856-1940) as the ultimate particles of electricity. An electron is an unimaginably small speck of matter and is perhaps the closest real thing to the cherished concept of the point mass particle of Newtonian physics. To this date, the electron is the smallest and lightest of all known elementary particles that have mass. We have not yet been able to put a definite size to it, but only infer from experiments an upper limit to its size.

While a firestorm raged all around it with the emergence of the nascent theory of quanta, the electron itself remained more or less untouched by the swirling controversy over duality. It was a particle, in the finest tradition of classical physics, and that was that. All that changed abruptly in 1924, nearly 30 years after its initial discovery, when a little-known French prince, Louis de Broglie, had the audacity to suggest that if radiation behaved like particles, then electrons should conversely behave like waves. This hypothesis was dramatically confirmed only 3 years later. Let's trace these two discoveries of electrons, first in 1897 as a particle and then in 1927 as a wave.

Electrons, the jacks of all trades

The electron is perhaps the most remarkable unremarkable particle that we have come to know. At least a half dozen superlatives are needed to describe the fundamentality of this little fella. Electrons were first discovered by J.J. Thomson as the basic carriers of electricity when in 1897 he made an accurate measurement of the ratio of charges to mass of particles that made up what had been called cathode rays. The first precise determination of the charge of the electron was made in 1911 by Robert Millikan. Combined with Thomson's ratio, this measurement yielded precise information on the mass of an electron. This was the same Millikan, incidentally, who confirmed the photon theory of Einstein in 1916. Preceding the discovery of the quanta of light by three years, the electron was the first elementary particle to be discovered. What were once considered to be elementary particles actually turned out later to be composite systems. The definition of so-called "elementarity" in particles has gone through several changes since then, but for almost the century of our acquaintance with it, the electron remains a truly elementary object to this date.

That an electron is an exceedingly tiny bit of matter is well evidenced by its mass, which weighs in at about 10^{-30} kilograms, or, if you like, about 2×10^{-30} pounds. The scientific notation of the powers of 10 is certainly convenient; it can hide our utter incapability of comprehending too many zeros. When there are 29 zeros after a decimal point, as in the case of the electron's mass, our minds simply go blank. Let's try an analogy instead. The mass of an electron is to the mass of a mosquito what the mass of a mosquito is to the mass of the sun.

An electron also has an exceedingly small size. So far as can be experimentally determined, it has practically no discernible size, the latest data registering its size to be less than 10^{-18} meters, that is, one-millionth of a trillionth of a meter. Now what is meant by the size of an electron perhaps needs some elaboration here. After all, it is not something that we measure with a desktop ruler. The "size" of an electron is defined by the furthest extension of its electric charges as can best be determined experimentally; a beam of very fast high-energy electrons is hurled at a target electron and the shortest distance they can approach the target before being significantly deflected by the mutual electric force sets an upper limit to the size of an electron. What is small or large is, of course, a relative matter, but by the standards of our human-sized world, an electron may as well be a point particle. In another analogy, we can say that the size of an electron is to the size of a human being what the latter is to the size of the Milky Way galaxy.

In addition to the extreme values for its mass and size, one other measure of small value characterizes what an electron is: its electric charges. The unit for electric charges, another convenient human-sized unit adopted back in the eighteenth century, is called a coulomb. In terms of it, the electric charges on an electron are about 1.6×10^{-19} coulombs. By the convention for the relative signs for charges adopted long before electrons were discovered, the sign of the electronic charges is taken to be negative. A typical hand-held pocket calculator runs on a 1.5-volt watch battery and has a power consumption of , say, 0.0006 watts. This corresponds to the flow of about 2.5 trillion electrons passing through a point inside a calculator every millisecond (thousandth of a second).

The electric charge of a proton, the hydrogen atom nucleus, is equal in magnitude and opposite in sign to that of an electron. The magnitudes of the two charges are not merely approximately equal, but are precisely the same as each other, better than one part in one hundred million. Such exact cancellation between the negative and positive charges in an atom is just one example of what appears to be a very rigorous rule of nature. According to this rule, called the rule of charge quantization, the electric charges on all observed subatomic "elementary" particles occur only in whole number multiples, positive as well as negative, of the charges of an electron. If we denote the magnitude of the charges of an electron by the notation q, the electric charges of all particles occur only in units of q, that is, zero, plus or minus q, plus or minus 2q, plus or minus 3q, and so on.

We do not know exactly what an electric charge is , any more than we understand what a mass is, but whatever it is and however it came into being what it is, this rule of quantization is obeyed by all the subatomic particles detected to this date. In this sense, an electron, or more precisely, the charge of an electron, acts as the "quantum" of electric charges much in the way a photon is the quantum of radiation energy.

This rule of charge quantization seems to be grossly violated by the picture of "elementary" particles in which protons and neutrons are themselves compound systems made up out of particles called quarks. In this theory, a proton and a neutron are two different combinations of two species of quarks called the "up" and "down" quarks. Two "up" quarks and a "down" quark make up a proton, and a neutron consist of one "up" quark and two "down" quarks. The electric charges of these quarks are positive two-thirds and negative one-third of q, respectively, for "up" and "down"quarks, the difference between them still being one whole unit of q. These fractional values, $+2/3$ and $-1/3$, are in clear violation of the rule of charge quantization, except for one thing. For almost 30 years since the introduction of the idea of quarks, no one has yet been able to actually observe and detect a single isolated quark. That is, no one has been able to do with a quark what Thomson and Millikan did with an electron, measuring its charges, or mass, or even its charge-to-mass ratio. For this reason, we cannot yet include quarks among the ranks of the observed particles.

Since its discovery in 1897, the electron had been the ultimate embodiment of the concept of a point-mass particle as defined in Newtonian mechanics. It held this lofty position for 30 years until 1927, when two independent experiments, one in the United States and the other in Britain, dramatically uncovered the hidden wave properties of it, thus verifying the hypothesis of a matter wave advanced by de Broglie. The British experiment was performed by George Thomson, the son of J.J. Thomson, who discovered electrons in the first place.

De Broglie's thing

In 1924, a year after Millikan received the Nobel prize in physics for having conclusively verified the photon theory of light, a doctoral candidate in physics at the University of Paris by the name of Louis de Broglie conceived an idea that was as crazy as it was bold: If a beam of light, a classical wave, should turn out to possess the dual characteristic of a wave and a particle, should it not also be true for a classical particle? Perhaps the electron, the very embodiment of a classical particle, might display some wave-like behavior if examined carefully within the small dimensions of the quantum world. Not only did he propose this new form of a wave for all particles, but he did it purely on the basis of what he perceived as the symmetry and reciprocity of nature, in a total absence of any experimental fact to support the claim. As de Broglie himself at first noted, his hypothesis was just a ''formal scheme whose physical content is not yet determined.''

De Broglie's proposal, however, did contain one important quantitative prediction that could be tested experimentally, either to be confirmed or discarded. Starting from the Planck-Einstein formula that relates the frequency to the energy of a photon, de Broglie came up with a formula that assigned a unique value to the wavelength of this new wave for every value of the momentum of the particle. It was an inverse relationship, i.e., the greater the momentum, the shorter the wavelength. In any case, the momentum of a particle is a readily calculable number and so is the wavelength of this new wave. De Broglie's hypothesis was something that could be put to test easily, either thumbs up or down.

As discussed in chapter 2, two physical phenomena are unmistakable signatures of a wave: diffraction and the interference. If something diffracts or interferes, just once, then that something is unquestionably a wave. Furthermore, the occurrence of the constructive and destructive interferences, the double-or-nothing trick, provides a quick way to determine the wavelength of a wave. So, de Broglie says that electrons are also waves? Okay. Shoot a beam of electrons into an appropriate setup and see if they diffract or interfere. If they don't, then de Broglie is wrong. If they do, go ahead and measure the wavelength and see if it agrees with what de Broglie predicted. An open and shut case. You can guess the rest. Three years later, in 1927, two experiments were carried out independently, and they both confirmed dramatically not only the wave nature of electrons but also the value of its wavelength exactly as de Broglie proposed in his formula. In 1929 a Nobel prize for physics was awarded to de Broglie, who remains to this date the only person to have received a Nobel for a doctoral dissertation!

The experiment that uncovered the wave nature of electrons is similar, in principle, to the ordinary diffraction and interference experiments involving X-rays and visible light. Let me first paraphrase the experiment, in terms of a more familiar setting involving light and semitransparent glass sheets. Suppose we have a stack of regularly spaced semitransparent sheets as shown in FIG. 4-1 and consider what happens when the lights reflected at different levels come together. Let us remind ourselves of the simple mechanism that produces the double-or-nothing trick, the constructive and destructive interferences discussed in chapter 2; when two identical waves are out of step by one full wavelength—or, for that matter, by two, three, or ten full wave-

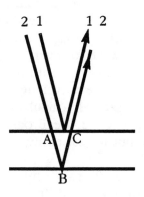

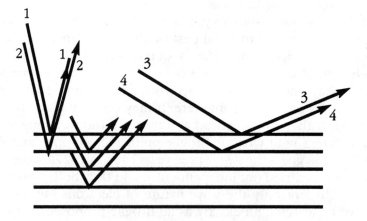

4-1 *A stack of semitransparent sheets.*

lengths—the two waves reinforce each other doubling their combined amplitudes. Further, when they are out of step by half steps, that is, half of a wavelength, or three halves, or five halves, they cancel each other. Since the sheets are semitransparent, some of the light that hits the top will be reflected and some will pass through. The portion that gets through the first sheet will repeat the split at the surface of the second sheet, some reflected back up while others pass through to the third sheet, and so on. Let us examine a pair of such successively reflected lights, reflected at the first and second sheets, respectively—the rays 1 and 2 at one angle of incidence and the rays 3 and 4 at a somewhat wider angle of slant.

Ray 2, reflecting back up from the second sheet, has clearly fallen behind ray 1, which is reflected at the top sheet. Ray 2 had to double back an extra distance, traveling from point A down to point B and back up from B to C. A similar situation exists for the pair of rays numbered 3 and 4. The difference in the distances traveled depends, in each case, not only on the width of the gap between the sheets, but also on the angle at which light strikes the surfaces. The difference in the distances traveled, or the path difference as it is called, is clearly greater between rays 3 and 4 than that between rays 1 and 2. And it is these path differences that bring out the interference effects. Whenever the path difference between a pair of such rays corresponds to a whole multiple of full wavelengths, we have a constructive interference, resulting in the doubling of the amplitudes and a bright reflection at that angle. Conversely, whenever the path difference corresponds to one half of a wavelength, we have a destructive interference, resulting in a complete cancellation of amplitudes and no reflection at all at that angle. By looking at the setup at different angles, or by rotating the whole stack of sheets against a sharp beam of light incident upon it in one direction, we observe a pattern of alternating bright and dark reflections.

An assembly of semitransparent glass sheets stacked up on a tabletop, while serving the purpose of illustration, is simply too crude for an actual experimental setup. The wavelength of light is on the average about 6,000 angstroms, or six-tenths of one micron. In order to produce a sharp interference pattern, the gap between the reflecting planes needs to be a few wavelengths wide. The wavelength of a typical X-ray used for this type of experiments ranges between 10 and 100 angstroms. We need a solid material the structure of which consists of orderly and precise geometrical patterns of tightly packed atoms and molecules. The precise lattice structure of crystals and some metals, therefore, provides an ideal setting with layers of atoms and molecules linked together serving as semireflecting planes in all three dimensions over a wide range of angles of incidence. A typical rectangular lattice showing a set of parallel reflecting planes, rich in atoms, is shown schematically in FIG. 4-2. This technique of shooting X-rays onto a crystal structure is how scientists investigate the structure of many solids. They use the interference pattern as the clue and work backwards to the structure, in what is a basic tool for the field of crystallography. This is also how we learned the structure of some biomolecules, such as the discovery of the structure of a double helix of a DNA molecule.

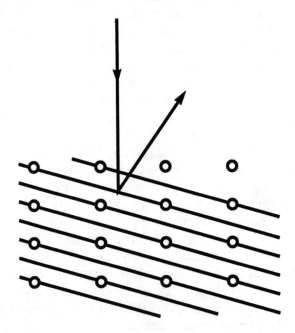

4-2   *Parallel reflecting planes of atoms in a lattice structure.*

By now, you can easily guess what must have transpired in the two experiments that confirmed the wave theory of electrons. In 1927 Clinton Davisson and Lester Germer in the United States and, independently, George Thomson in Britain, performed experiments in which they shot a beam of electrons into foils of various metals and observed the reflections of the electrons at different angles of incidence. Much to their astonishment, they observed—yes, that is right—an unmistakable pattern of interferences, an alternating pattern of maxima and minima in the number of reflected electrons. Whatever else they might also be, the beam of electrons behaved just like a wave and with the wavelength that de Broglie predicted it would have. Davisson and Thomson, but not Germer, were awarded the Nobel prize in physics in 1937.

This newly discovered wave of electrons was an entirely new kind of wave. Unlike the electromagnetic wave, or the wave of photons, the new wave, which was called the de Broglie wave of the matter wave, existed only within the realm of the quantum world. This fact perhaps contributed to a deeper difficulty in developing an intuitive feeling for it and for coming to grasp the sense of the wave-particle duality. Nevertheless, this matter wave is exactly what brings the picture of duality to its completion. The relationship between the particle aspect and the wave aspect is exactly the same for matter as it is for radiation. By 1927 the dual nature of matter as well as radiation had been firmly established by experimental facts. What happened next is the conceptual and philosophical interpretation of this strange phenomenon, to which we now turn in the next chapter.

5

✳

The heart of quantum mechanics

SO, AFTER ALMOST THREE DECADES of one startling development after another on both the experimental and theoretical fronts, the situation around 1927 was anything but clear. All the pieces of the puzzle had been uncovered, but they had yet to be put together. Experimental results left no room for doubt that when radiation interacted with individual atoms and electrons, it behaved like a stream of lumpy and localized units of discrete energy and momentum, namely, particles. And when a beam of electrons displayed the diffraction and interference patterns, there was no denying that it was a wave.

What was once a hallmark of a classical wave turned out to be particles, and, conversely, the archexample of classical particles was shown to harbor wave-like properties. The distinction between the two had completely disappeared: Both radiation and matter were somehow both a wave and a particle at the same time. The blatant self-contradiction, at least by the standard of classical physics, of the whole situation was too obvious, and the founding fathers of quantum mechanics had to struggle with it for all that time. It isn't going to be any easier today, but let us delve right into the thick of it, because delve into it we must now do.

The dance of duality

In chapter 2 I laid out the essence of the distinction between the concepts of a particle and a wave as defined by classical physics, not only what each is but also how each behaves characteristically. Under appropriate conditions, such as being incident upon a small hole or when more than one wave are superimposed, the waves diffract or interfere. No Newtonian particle can do that. Particles, on the other hand, have their own unique way of interacting with each other. They collide at a point, like billiard balls, so that one particle strikes the other, scatters, and recoils, transferring energy and momentum at a localized point. In our human-sized world, these two physical attributes never overlap; one is one and the other is the other, and that is all there is to it.

Much to our helpless befuddlement, however, what happens in the world of quanta is exactly contrary to such a clean distinction. The same physical entity, be it an electron or a photon, can and does exhibit both the wave and particle properties. Under some circumstance, it diffracts and interferes and, with other particles or individual atoms, it interacts as a particle. When a beam of photons, that is, a beam of light, is directed toward atoms, an individual photon may strike an electron head-on or miss completely. When a one-on-one collision occurs, the photon will scatter off in one direction as the electron recoils into another direction, just like a couple of billiard balls would do in a collision. Conversely, a beam of electrons incident upon a small hole will diffract—bend around the corner and spread out in all directions—in complete agreement with the wave hypothesis of de Broglie.

Actual experiments that reveal this dual characteristic of nature are carried out in intricate setups that can measure minute dimensionalities of the quantum world. It is a little difficult for us, without the benefit of having an actual example in our human-sized world, to develop a genuine feel for this deep mystery. Instead, we try to transcribe the situation up to our scale of things by cooking up deliberate and hypothetical analogies which, while not wholly accurate, serve the purpose of illustrating the essence of the wave-particle duality. One such analogy involves the diffraction of a wave through a small hole, as illustrated in FIG. 2-5 of chapter 2, in which we showed what exactly particles and waves do—and do not do—when incident upon a small opening.

Let's examine what would happen in a simple diffraction experiment in which a beam of things, matter or a wave, is directed toward a very small opening, beyond which is set up a suitable screen to detect the arrival of whatever passed through the hole. Such a setup is schematically shown in FIG. 5-1. In an actual experiment, the beam of things hurled toward an opening would be a beam of either electrons or photons, the width of the opening should be on the order of the wavelength of these particles, and the screen would be either a sensitive photographic plate or a battery of electron detectors. But since we are pretending to have an analogy in our human-sized world, the screen can be imagined to be a wall of Sheetrock, where flying pellets would make pock marks, a cardboard with a sticky surface to catch grains of sands blown in, or, for that matter, a piece of cloth to catch an incoming breeze.

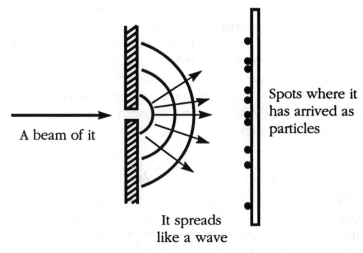

It spreads
like a wave

5-1   *The dance of duality.*

Let's further pretend that a steady stream of small pellets is aimed at the hole. As some of the pellets emerge from the hole to the other side, we are observing the pattern of their arrival at the screen. Something totally inexplicable and utterly incredible would happen: The pellets would strike the screen all over the place, up and down and sideways, as shown in FIG. 5-1, seemingly at random. The pellets have apparently bent around at the opening and spread out in all directions, just as any self-respecting wave would and no self-respecting Newtonian particle ever could. Let us try the experiment the other way around. Suppose we send in a wave, that is, what we think is a wave, through the hole and see what happens at the screen. As a wave, it would most certainly diffract and propagate out in all directions from the hole toward the screen. Rather than the arrival of a continuous and fluctuating wavefront, however, what we would observe at the screen is a pattern of discrete marks made by the striking particles again all over the place. In both cases, what is observed at the screen are the traces of individual particles, but the pattern of these marks could only have been possible by the diffraction of a wave at the hole.

We have described the dumbfounding mystery of the wave-particle duality in terms of a deliberately hypothetical analogy, something in terms of the scale of our world. But every actual experiment using electrons and photons, in a setting similar to what we have just described, yields the same result, unmistakably displaying the characteristics of both a wave and a particle. It must somehow then be both: It spreads like a wave but impacts like a particle. That is the essence of the duality. However paradoxical it might appear to us, it is exactly what occurs in the realm of quanta. This is the new quantum reality and the very essence of it—the heart and the soul of quantum mechanics—exists between the hole and the screen.

The path of probabilities

The essence of the wave-particle duality—that it spreads like a wave but impacts like a particle—can easily be misunderstood, as is indeed sometimes the case. One common misunderstanding is to view the wave as a collective manifestation of a multitude of particles, thousands or millions of particles forming a wave. It is not difficult to fall into such a trap. A beam of ordinary light, after all, contains zillions upon zillions of photons, and billions upon billions of electrons form a beam of electrons used inside an electron microscope. All classical mechanical waves are this way. Water waves, if examined at a molecular level, consist of billions of individual water molecules. The alternating patches of compressions and rarefactions of air molecules are what we call a sound wave. The wave in the quantum mechanical wave-particle duality clearly cannot be a wave of this type. If it were, it wouldn't have caused the kind of conceptual revolution that it has.

The wave-particle duality of the quantum world is a totally new reality of nature, the likes of which cannot be found in the dimension of things in our human-sized world. The "waveness" and the "particleness" coexist and maintain this duality all the way down to the behavior of a single physical entity. Right down to a single photon or a single electron, it is a wave and a particle at the same time. This is the quantum duality, and this, at least by the standards of classical physics, is the paradox. Again using the single-hole diffraction setup, let's illustrate this bizarre reality.

As sketched in FIG. 5-2, suppose we conduct an experiment in which a bunch of electrons, say nine of them, are shot through a small hole and detected at the screen a short distance away. They are shot into the hole one electron at a time with enough separation in time to allow for individual detection at the screen. Experiments similar to this setup have been performed as recently as 1961, and what is being described in the figure corresponds to a simpler paraphrasing of such an experiment. Needless to say, any consideration of any electron hitting the sides of the opening and scattering off is not allowed. Under ideal conditions each of the electrons goes right through the hole. According to Newtonian mechanics, there is absolutely no way in which the particles can, by themselves, change the direction of their motion and bend their path. Consequently, they would have all landed on the same

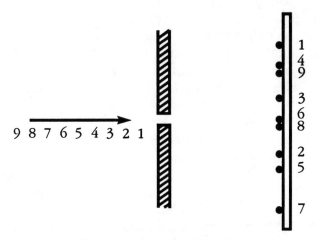

5-2 *The riddle of duality.*

spot on the screen directly opposite to the hole, that is, nine straight bull's-eyes. Well, that is not what happens.

What actually takes place is something wild and crazy. Electron 1, the first electron to go through the hole, is detected at the position marked 1 on the screen. It certainly has diffracted. Another electron is shot through the hole, and it lands at a spot marked 2. The spot where the third electron hits the screen is marked 3, and so on. Every electron has been detected on the screen as individual localized particles, yet at the same time they all have diffracted at the hole. In passing through the small opening, the "waveness" has prevailed, but when observed on the screen it is a particle that struck a spot. Furthermore, one electron went to one spot and another, under an identical condition, went to another spot. Is the diffraction of an electron a completely random process, or is some hidden rule operating here? If there is such a rule, what might it be? Questions, befuddlements, and more questions. How are the two aspects, the "waveness" and the "particleness," to be resolved, and how are they related to each other? These questions go straight to the heart of quantum mechanics, and the answers to them define the bedrock of quantum theory.

The resolution of this apparent paradox, the simultaneous coexistence of two contradictory attributes, did not come about easily. When it was gradually and finally developed through much thought and long deliberations, especially by Niels Bohr, Max Born, and Werner Heisenberg, the interpretation of duality required a totally new and revolutionary viewpoint of physical reality. The culmination of this resolution is usually referred to as the principle of probabilistic interpretation. According to this principle, the "waveness" and "particleness" are related to each other by a mathematical recipe for a probability distribution. In other words, a quantum physical entity, be it a photon, electron, proton, or atom, is a wave, and its shape controls the probability it will be observed at a particular position. Suppose a particular wave has the shape shown in the upper diagram of FIG. 5-3, the wave amplitude being positive as well as negative. This amplitude harbors the following precise information: At every position the amplitude squared, that is, the height or depth of the wave at that point multiplied by itself, and hence always positive, corresponds to the probability with which the quantum entity will be found—detected or observed—at that point. The higher the amplitude

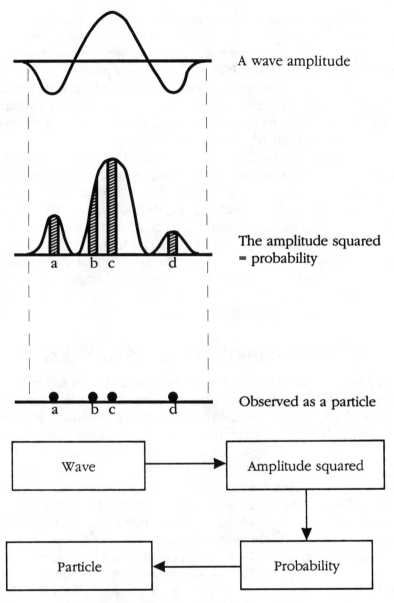

A wave amplitude

The amplitude squared = probability

Observed as a particle

5-3 *The probability connection.*

squared, the greater likelihood it will be found at that position. In FIG. 5-3, the object has the highest probability to be found at or near position c, much less at position d, and zero chance to be found near the two valleys where the amplitude squared drops to zero.

It is truly a strange admixture, if it can be called that, of the dichotomy of a wave and a particle; it is neither a wave *per se* nor a proper particle. It is a wave that contains precise information of probabilities with which it can be seen as a particle and, at the same time, it is a particle that with its endowed wave properties will diffract and interfere. There you have it, all of it, the probabilistic interpretation of the wave-particle duality which is the very defining foundation of all quantum physics.

As I mentioned in chapter 1, this idea that in the end, at the level of the world of quanta, there is no completely determinable objective physical reality out there and that everything is governed by some fluffy laws of chance, did not sit very well, to say the least, with some founding fathers of quantum mechanics. These included not only Albert Einstein and Louis de Broglie, but also Erwin Schrödinger, who discovered the central mathematical equation of quantum mechanics called Schrödinger's equation. Quantum mechanics as we know it today, however, has stood on this foundation for more than 90 years.

The certainty of uncertainties

The first consequence of the wave-particle duality is self-evident: The complete determinism of a point-mass particle occupying one position in space at one moment in time automatically goes out the window. The new quantum reality is such that a matter wave carries with it the information—the matter wave amplitude squared—on the likelihood for it to be found as a particle at various locations. Since a wave, any wave, has to have some spread over a distance, however short, the position of a particle at any given moment cannot have a single definite value but is probabilistically spread over the wave's extension.

Again referring to FIG. 5-3, suppose that at a given moment in time, an electron wave has the shape shown in the figure, as well as the amplitude squared. As far as the measurements of positions are concerned, what is certain is where we are not going to find it: at or near the two dips of the amplitudes squared and, of course, to the left and right of the ends of the wave's extension. What is likewise certain is that the electron will be found at some point in between. We can never be certain, however, at exactly what position we will find it; that is, we are certain of the inherently binding uncertainties in the outcome of the measurements of an electron's position. The most likely spot is at or near position c, less likely at position b, and even less so at positions a and d, for example. Let us say that the probability at position d is about 3 percent. All we can say is that in 3 out of 100 measurements, the electron will be found in that position, and 97 times it will be found elsewhere. The position of a particle is inherently uncertain because of the wave-particle duality and its probabilistic interpretation, and in this sense the uncertainties in the quantum world are direct consequences of the duality.

Often, and quite correctly, the quantum mechanical uncertainties are introduced as the consequences of an act of measurement. The very physical act of measurements—weighing, clocking, sizing, or, for that matter, touching, sniffing, or just plain looking—do not disturb, in our human-sized world, the object being measured, certainly not to the extent that it interferes with the expected result. In the quantum world, where the magnitudes of physical entities being measured and the agents of measurements are comparable to each other, however, it is a quite different situation concerning the act of measurement. A photon bouncing off an electron can and does create disturbances not unlike one car crashing into another, forever altering the state of that electron. A second photon bouncing off the same electron will do so not only at a different location, but also while the electron is in a completely different state of motion. The resulting uncertainties are totally unavoidable. This is how, in one picture of quantum mechanics at least, an observer becomes a part of what is being observed.

Either explanation of the uncertainties is just as good as the other. Insofar as we must deal with what is actually observed and describe nature in terms of the results of physical measurements, either explanation will lead to the same set of uncertainties. The interpretation in terms of the disturbances by the acts of measurement, however, does tend to leave behind a nagging doubt whether a particle, in principle, could have had an exact and unique position before the measurement. A state of matter in the absence of measurement is, of course, quite meaningless. The uncertainties as a consequence of the probabilistic interpretation of the wave-particle duality, on the other hand, clearly show their inherent nature and are built into the very definition of physical objects.

6

A tunnel
without digging

So, AS I HAVE JUST ELABORATED in the previous chapter, the
way in which we define, measure, and analyze the physical world
surrounding us went through a totally revolutionary overhaul in
the first three decades of the twentieth century. At the end of the
30-year journey down a winding path through the new identity of
the wave-particle dualism—the probabilistic interpretation of it
and the inherent uncertainties in measurements that it entails—
we have come upon a completely new theory of matter: the wave
theory. This new theory is built on the fundamental premise that
all material specks—molecules, atoms, electrons, atomic nuclei,
protons, neutrons, and many more yet to come—are in the end
both particles and waves, correlated to each other through the
rule of probability.

Now this business of a physical thing behaving sometimes as
a wave and sometimes as a particle is not something we can visualize and perceive, easily or otherwise. All our concepts and intui-
tions have been developed on the basis of the classical description of the nature. Daily experiences constantly reinforce
this Newtonian view of the physical world. As contradictory as it
appears to us, however, we have little choice but to accept it in
full for one compelling reason: Experiment after experiment have
confirmed this dual characteristic to be an unassailable basic
property of the quantum world.

It is interesting to observe, however, just how tenaciously—and perhaps desperately—we cling to the habit of using the name particles. Subatomic species are always referred to as the subatomic particles: after all these years, some 90 years since the dawn of quantum physics, no one ever invokes the name subatomic waves. Electrons will always be the particles of electricity and atomic nuclei will always be made of nuclear particles, the protons and neutrons, never of nuclear waves. It is difficult enough to rally support for the proposed superconducting supercollider (SSC) as the largest elementary particle accelerator in the world. With its estimated cost escalating daily, topping the $10 billion mark, no one dares to refer to the SSC as the largest elementary wave accelerator! The name particle is definitely a poor substitute for the wave-particle duality, but it is evidently a substitute vastly preferred to the other alternative, the wave.

The wave theory of matter, as we know it today, firmed into an exact mathematical formalism two years after de Broglie came up with his original idea, when, in 1926, Erwin Schrödinger (1887 – 1961) worked out the famous equation bearing his name. A theory for any type of waves requires a defining equation that governs the behavior of the waves under all possible conditions, and Schrödinger wrote down just such an equation for matter waves. It had to satisfy many requirements: It not only had to encompass all three founding principles, those of duality, probability, and uncertainty, but it had to provide all information about a matter wave for all time and under all physical circumstances. The equation of Schrödinger did all that, and a lot more, and became the mathematical centerpiece of quantum mechanics.

As scientists probed deep into the microcosm, the land of duality, and began to unlock the inner secrets of matter, a whole new world of matter emerged, a world full of all sorts of strange and bizarre phenomena, totally unexplicable by the standards of the Newtonian concepts. One particularly bizarre happening involves a situation in which a particle—oops, a wave, no a duality—such as an electron or proton, displays an uncanny ability to penetrate an impenetrable barrier, that is, a physical barrier that is absolutely impenetrable by the rules of classical physics. This little-known phenomenon in the backwater of quantum physics, called quantum tunneling, has leapt out of relative obscurity and, beginning in the 1980s, burst onto the center stage of the latest high technologies in microelectronics, microscopy, and quantum material science. In this chapter, after a brief nonmathematical description of what Schrödinger's equation is all about and what it has done, I focus on this strange occurrence of quantum tunneling.

The equation of Schrödinger

The questions we ask in order to physically describe a matter wave have to be quite different from those for a macroscopic material point. For the latter we need answers to such questions as its precise position at any given instant of time, its speed and direction, its rate of acceleration under the influence of a force, the overall trajectory of its motion, and so on. These questions have become, completely meaningless in the world where identities became blurred in the wave-particle duality and movements are clouded by uncertainties and chances of probability. The rules of the game have changed, and so have the questions. Instead, we ask: What is the shape of a matter wave? Or, more precisely, what determines its shape at any given instant of time? How does it move—or wiggle—and propagate itself? What happens to the matter wave when a certain force is applied to it? What is its spread end-to-end? As mentioned before, when de Broglie first proposed the idea, he did it not only in the total absence of any experimental facts to back it up, but also without any concrete mathematical framework in which answers to these questions could be sought. It was indeed a wave in search of its equation.

The equation in question had to perform a formidable task: It had to be able to make precise sense out of the world of blurred identities and uncertain movements. It had to be an exact equation that could calculate and predict how a set of probabilities change into another set of probabilities in time. Not unlike the challenge of weather forecasting, it had to be able to predict the flow of patterns of clouds and winds—except the clouds and winds are matter waves in the land of dualities.

As a bunch of electrons traveled in different orbits around an atomic nucleus, some even switching orbits as they went, we had to know the shapes of electron waves completely for all time. As electrons jumped tracks from one atom to other nearby ones, locking atoms together into molecules, the equation had to tell us how the electron waves managed to pull off these tricks. To answer all these questions and a lot more, and to claim the position of the mathematical centerpiece for the new wave theory, the equation had to be, well, 10 feet tall, sharing ranks with the equations of Newton, Einstein, and Maxwell. Such an equation Erwin Schrödinger found in a remarkably short order, within a span of less than six months in 1926.

Viewed purely for its mathematical structure, the wave equation of Schrödinger is what is technically known as a partial differential equation for an eigenvalue problem, certainly an imposing mouthful of a name. Compared to other well-known equations of physics, such as those of Newton, Einstein, and Maxwell, a lot more mathematical skill is required to handle it. The input information to the equation is the force influencing the dynamics of a particle—the absence or presence and nature of the force. The output is the solution, a mathematical expression that spells out the shape of the matter wave in question and how it changes in time. This sounds simple enough, but in most cases it takes a lot of doing before we can squeeze out a solution.

A proton might be cruising along freely in the absence of any force, minding its own business. A passing electron might be attracted to a proton by the pull of the mutual electrical force and find itself going around in an orbit centered at the proton, forming what we call a hydrogen atom. Or a whole bunch of electrons might be "shooting the rapids," skillfully negotiating a path through a maze of atoms inside a copper wire, forming what we call an electric current. The procedure for obtaining the form of matter waves is simple: Specify the force in question, no force at all, a single source of force or a lattice of multiple sources, and solve Schrödinger's equation in one way or another.

A unique feature of Schrödinger's equation is that the shape of a solution is keyed to it energy, a different shape for a different value of energy the matter wave has. In FIG. 6-1 are some typical examples of matter waves corresponding to a particle moving along freely in the absence of any interfering force, on a straight line and with a constant speed. Generally, the faster the speed of a particle, that is, the higher its energy, the shorter its wavelength.

Called a wave pulse or a wave packet, the types of waves shown in FIG. 6-1—that is, "snapshots" of waves at one instant of time—correspond to some degree of localization for the particle it represents, the probability of location confined within the spread of the waves. In other words, the inherent uncertainty regarding the position of a particle is represented by the spread of

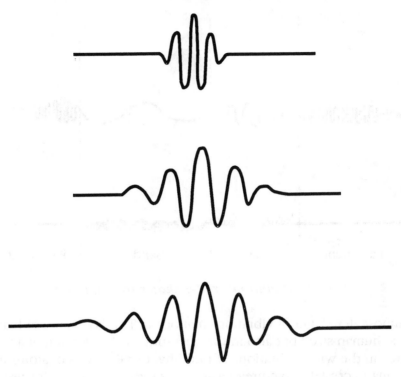

6-1 *Examples of matter wave pulses.*

its wave. Another example is a moving particle that runs into abrupt changes in forces that speed or slow it. Suppose we have a roller skater who, while coasting down a smooth pavement, runs into a patch of grass or, worse, a sand pit. Provided he was small enough to be a "quantum" roller skater, his quantum wave would go through the contraction and re-expansion of its wavelength, something like shown in FIG. 6-2.

Things become more interesting, and more complex mathematically, when a force is strong enough to confine the motion of a particle within a region of space, binding the particle to the source of the force—high walls boxing in a particle, electrons going around in orbits bound to an atomic nucleus, or a particle oscillating to and fro attached to an end of a spring. Let's take the case of a spring as an example. No one is suggesting that we need an actual spring small enough to attach an electron to it. The oscillatory motion associated with a spring draws its universal importance from the fact that it represents the basic mode of motion for

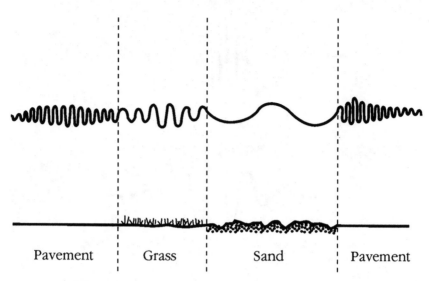

Pavement | Grass | Sand | Pavement

6-2 *A "quantum" roller skater rolling along.*

almost all manner of vibrations that occur. This is true not only in the human-sized but also in the quantum world: swaying of a tall tree in the wind, vibrations of pudding, or vibrations of atoms in a quartz crystal. A picture of a spring is the simplest way for us to visualize an oscillatory motion.

Consider the simple setup depicted in FIG. 6-3. A block is attached to a spring which is firmly anchored to a wall. When not disturbed the block and spring just sit in the normal position, the unstretched and uncompressed position, marked 0 as shown in diagrams (a) and (b). Suppose we pull the block out to the position marked +A and let go. The block will execute its motion, as every child knows: It speeds up, passes by the original position with maximum speed, slows to a momentary stop at the other end marked −A, reverses its direction to repeat it again and again, back and forth.

Let's first examine what a simple Newtonian analysis of the old-fashioned classical probability would look like for this block. It is quite simple. The block spends the least amount of time at or near the midpoint, while most of the time it hangs around the ends: It speeds across the midpoint with its fastest speed while it coasts to a stop at the fullest extension before turning around. The chance of seeing the block at either end is therefore much greater than the chance of seeing it in the middle. The classical probability curve would look like the one in diagram (c).

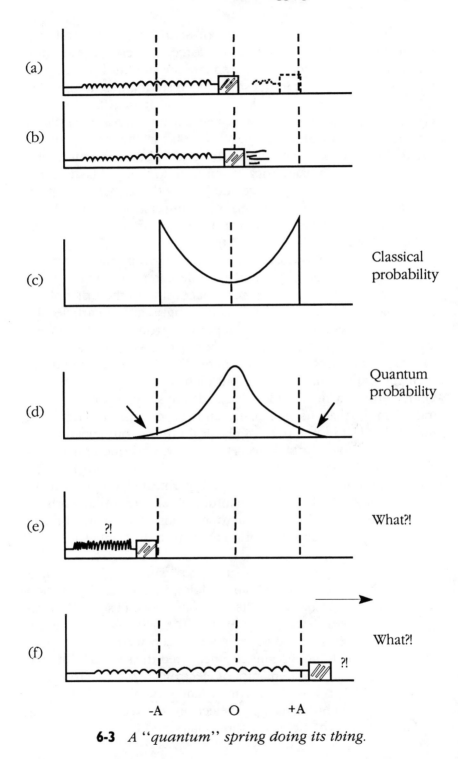

6-3 *A "quantum" spring doing its thing.*

When we solve the equation of Schrödinger for the block under the influence of a spring-like force, the result of the wave we get is entirely different. If we square the amplitude of the matter wave for the block in question, the resulting quantum probability looks like what is sketched in diagram (d). The stark contradiction between the classical and quantum results couldn't be clearer. The quantum probability peaks in the middle, meaning that the block is most likely to be found near the midpoint. This conclusion is diametrically opposite to that of classical mechanics. An even more strange and bizarre thing about the quantum wave has to do with the gently sloping tails of the probability curve that actually extend beyond the maximum extension, as indicated by arrows in diagram (d).

The quantum probability, instead of dropping to zero at regions beyond the maximum possible extension, actually juts out further before becoming zero. According to the principle of probability, this can mean only one thing, that the particle, the block in this example, can exist at points beyond the position $+A$ and likewise at points below the position $-A$. Chances are not that great, but it can be found there. You have a block and the spring. You pull out the block four inches before you let go of it. On its return swing, the block extends the spring six inches, two inches more than its maximally allowed extension! This is an impossible feat by the rules of classical physics, but apparently a piece of cake for a matter wave. How can we describe a phenomenon that can never take place on our side of the world? How do you say it: jutted out, leaked, seeped, permeated, crept through, or penetrated? As a wave, a quantum object can do all sorts of strange things, and this is but one example of many bizarre goings-on in the quantum world. The world of dualities is full of mind-bafflers.

Despite many instances of such perplexities and what at first appear to be downright impossible riddles, Schrödinger's equation delivered magnificent results: Its solutions under all sorts of physical circumstances described accurately the behavior of matter at the quantum level. The agreements between theory and experiment were simply impeccable. It uncovered the secrets of atomic structure, the intricate dynamics of electrons revolving around an atomic nucleus in a well-prescribed family of orbits. How molecules are formed by exchanges of electrons between nearby atoms and how millions of molecules form a latticework of military precision, turning themselves into useful solids for us became clear by the works of this equation. This newly gained knowledge formed the basis of our quantum mechanical understanding of the mechanical, electric, magnetic, thermal, optical, chemical, and bio-chemical properties of matter. All of atomic, molecular, and solid-state physics, in fact, have been built on this single equation. With the exception of the behavior of elementary particles at very high energies, that is, at energies higher than those needed to form atoms and atomic nuclei, what we mean by quantum physics boils down to the task of solving the Schrödinger equation in one way or another, either completely or, if it has to be, by a reasonable approximation.

And making a leap from the base of new knowledge of atoms and molecules led to the dazzling new technologies of this century, the so-called high technologies of today: The discovery of artificial semiconductors that led to the invention of transistors, which in turn ushered in the age of information, the age of computers, and instant global telecommunications, as well as the discoveries of lasers, superconductors, and biomolecules. Schrödinger's equation, the bread-and-butter workhorse of quantum physics, for nearly 70 years now since its original formulation, is yet to turn in another stellar performance leading us into still newer and more breathtaking technologies of tomorrow.

The art of quantum tunneling

The phenomenon of quantum tunneling, to which I have already alluded, is sometimes called barrier penetration. But neither name, tunneling or penetration, does full justice to what actually takes place. It simply can't. What occurs in the process has absolutely no analog in our human-sized world. Consequently, there is no word to adequately describe it. The tunneling effect is an exclusively quantum phenomenon that refers to the dumbfounding ability of particles to "penetrate" or "tunnel through" a physical barrier utterly impenetrable by all rules of Newtonian physics. And, to top it off, a particle is able to pull this trick—hold onto your seat—without ever actually having gone through the barrier!

For such a daring feat, the effect is actually one of the elementary consequences of Schrödinger's equation, much simpler in mathematical terms than, say, the systematic unraveling of the atomic structure. As we have already seen in the case of an oscillating system, a matter wave can and does behave in strange ways. The case of a particle bumping up against an unyielding barrier is another example of the mysterious workings of quantum mechanics. The experimental confirmation of the tunneling effect came swiftly: Less than two years after the work of Schrödinger, George Gamow, the originator of the idea of the Big Bang theory of the origin of the universe, applied this new effect in 1928 to solve one of the long-standing puzzles in a class of nuclear reaction. To this date, the tunneling theory stands as the only explanation, as I discuss in the next chapter, of a class of nuclear transmutations called the alpha particle emission, or alpha decay for short.

Any analogies of the tunneling that we can cook up in terms of things we can easily associate with cannot be expected to be realistic; things just do not happen that way in our world. Nevertheless, a couple of examples—desperate analogies—might serve to bring out the flavor of what this is all about. Sometimes we have to remind ourselves that we are, after all, talking about the goings-on in a world in which the size of a small molecule is considered to be very large.

Let's consider a ball rolling down and up a rollercoaster-like track, as shown in FIG. 6-4. Suppose the ball is momentarily held at and released from position A. It will quickly roll down the slope and climb up the hill toward the peak, position C. Of course, it can never scale the peak—the highest it can climb is

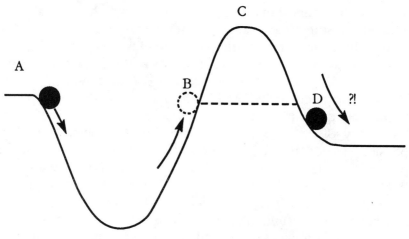

6-4   *A tunnel without digging.*

position B, which is at the same height as position A. The ball would go up and down and oscillate between the points A and B, without friction forever. There simply is no way in which the ball, released from position A, can ever be seen to be rolling down the other side passing through position D. At least not by the rules of Newtonian physics. But that is exactly what is observed in the quantum domain: The ball can roll down the other side of the slope, minding its own business! Not every time, but often enough. Does it actually tunnel through the peak along the dotted line? No. Does it go over the top? No. After climbing up to position B, the ball just materializes on the other side. That is the puzzle, the riddle called quantum tunneling.

Let's give it another go, another ridiculous analogy. One look at FIG. 6-5 should tell you what is coming, as surely as a ball coming through. Let's say we have two indoor handball courts adjacent to each other sharing a common reflecting wall. As players

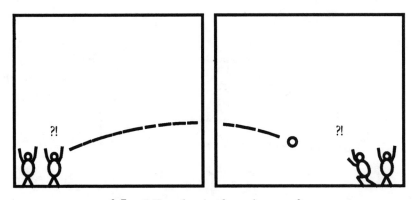

6-5 *OK, who is the wise guy?*

go at it, something incredible unfolds before their eyes: Once in a while a ball hit against the shared wall literally disappears right into it and shows up, out of nowhere, right out of the wall in the next court! The wall shows absolutely no damage and no evidence that anything has ever gone through it. It is like one of those dancing fogs that goes through a solid wall in a ghost movie, the dancing fog in this case being a quantum matter wave. Let us take stock of this befuddling situation in a slightly quantitative manner in the next section.

The science of it

The desperate examples I cooked up in order to illustrate the tunneling effect, a hill and a wall, are deliberate analogies. In the actual quantum world, physical barriers are not all that readily visualized. It might be a tiny region of an insulating medium that a flow of electrons encounters, or a strongly repelling force field, a sort of a strong "headwind" blowing against an electrically charged atom (an ion) as it approaches an electrically charged region. In order to discuss some quantitative aspects of tunneling, therefore, we need to set up a standard way of representing these barriers in a quantitative manner. To do so it is convenient to fall back, once again, on a familiar everyday situation. Let's recall the pavement-grass-sand combination of FIG. 6-2 and consider a golf ball rolling across it from left to right, a detail of which is shown in FIG. 6-6.

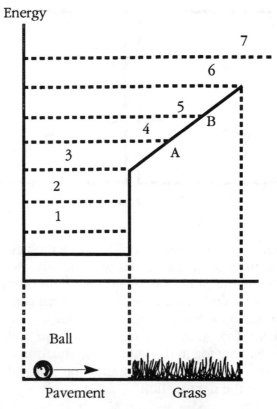

6-6 *Energy profile of a barrier.*

The pavement portion, assumed as usual to be a perfectly smooth and frictionless surface, provides the room for a running start for the ball. As it reaches the boundary between the pavement and grass, the ball will do one of three things. Either it comes to a gentle stop, it continues on to the grass area and covers some distance before coming to a stop in the grass, or it goes clear across the grass. Obviously the path of the ball depends on two factors: the strength of the motion of the ball and the strength with which the grass resists and impedes the motion—that is, the energy of the moving object—the ball—and the energy of the resistance by the barrier—the grass. To incorporate these energy considerations, it is convenient, and in fact, customary, to plot these energies on a vertical scale. This scale serves to indicate the levels of energy of the moving object and provide an energy profile of the physical barrier. The energy profile of the patch of grass is shown in FIG. 6-6, somewhat resembling the silhouette of a landmark skyscraper in Manhattan.

The two lowest energy levels, marked 1 and 2, correspond to a situation in which the ball's energy is too small for it to cross into the grass. The ball would come to a stop at the boundary, represented by a sheer vertical "wall." The highest point of the "wall" represents the minimum energy—the minimum speed— needed for the ball to cross into the grass, the energy corresponding to level 3. With a little higher energy, the ball rolls into the grass, as far as position A with energy level 4, or position B with a still-higher energy. A ball moving with energy higher than the amount indicated by level 6 would sail clear across the patch of the grass. The energy profile of a barrier is a convenient way of analyzing the situation. All right, you say, but what about the sand pit, the nightmare of golfers? That is easy. Without any further ado, I give you the energy profile of the whole situation, shown in FIG. 6-7.

The height of an energy profile thus has less to do with the actual height, if any, of a physical barrier, but represents instead the scale of energy required of an incoming object to overcome it. In this way of quantifying a barrier, the wall between the two handball courts of FIG. 6-5, which a handball cannot penetrate under normal circumstances, would be represented by an energy profile with a dizzying height—at least as high as shown in FIG. 6-8, a mile-high tower! Now all manner of barriers would be quantitatively represented, in principle, by all sorts of complex energy profiles. In most cases, however, the barriers we deal with in

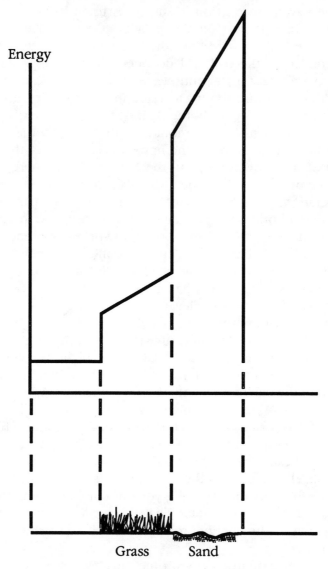

6-7   *Energy profile of a double barrier.*

quantum mechanics have a relatively simple bumpy profile, some of which are sketched in FIG. 6-9. The ones with straight lines and sharp angles lend themselves to easier and simpler analysis, but they are mathematical abstractions of an actual situation corresponding to gradual lines and rounded corners, as one would expect from a barrier of a repulsive force field, for example.

6-8 *Energy barrier of a solid wall.*

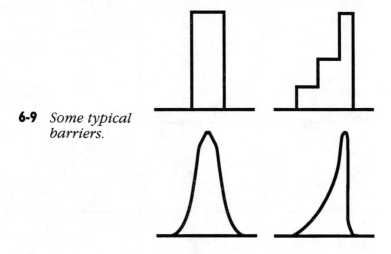

6-9 *Some typical barriers.*

Now we come to the crux of the matter, the phenomenon of quantum tunneling, a trick that a matter wave pulls off when it encounters an unscalable barrier. Suppose we have a matter wave, be it a tiny proton, or still tinier, an electron. Suppose it approaches and bounces back from a force barrier, be it an insulating medium inserted in the path of an electric current or a strongly repelling force region such as an electrically charged atom. Schematically, we have a matter wave of a certain energy approaching the energy profile of a barrier, such as shown in FIG. 6-10. Strictly speaking, the incoming and reflected waves are at the same height, but a little spacing is used between them here solely for the purpose of illustration. A wave wiggles in, hits the barrier, and, not having sufficient energy to overcome it, is reflected back out.

As the matter wave hits the barrier, owing to the very nature of a matter wave, what transpires is a little—well, a lot—different. As I elaborated in previous chapters, the matter wave stands for the probability distribution of locating the particle, and hence the position of the particle is inherently uncertain by the extension of the wave. We can only say that the particle is somewhere within the size of the wave near the point where the wave and the barrier come into contact. The matter wave splashes against the barrier, gathers itself and returns, and all the while the exact location of the particle is uncertain within the extension of the matter wave.

A detailed explanation of exactly what happens to a matter wave as it splashes against a barrier is shown in FIG. 6-11. As a general rule of the equation of Schrödinger, a matter wave must maintain its form without sharp changes. This means that, among

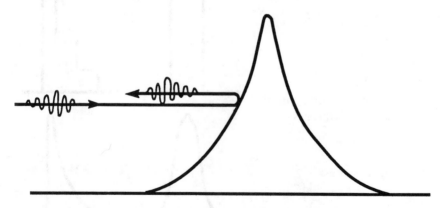

6-10 *A matter wave bouncing off a barrier.*

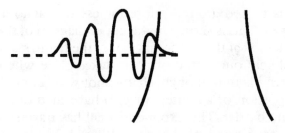

A wave protrudes into the barrier a little, but is
reflected at a thicker portion.

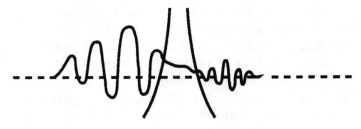

Where the barrier is thin enough, it
resurrects itself at the other side.
It has tunneled through!

6-11 *The trick of a matter wave.*

other things, when the front edge of a matter wave comes to a
barrier, it will not simply bend its shape just to conform to the
contour of the barrier. Instead, the front edge of the wave must
protrude a little into the forbidden zone. What is energetically
illegal by the rules of Newtonian physics is likewise energetically
illegal in the quantum domain also, but in the latter case a little
violation is tolerated as long as it is within the bounds of the
inherent uncertainties that go with a matter wave. In this way, a
nonzero probability in the forbidden zone becomes possible.
With this intrusion of a matter wave into the forbidden zone,
either of two possibilities takes place.

When the barrier is wide enough, that is, wider than the protruding portion of a matter wave, nothing dramatic happens; the
probability inside the barrier dies out rather quickly—exponentially, to be exact—and other than the slight intrusion, the matter
wave is reflected at the boundary. An electron approaching a
strongly repelling force field goes as near as it can, consistent with

its energy, and is turned around, more or less similar to a ball bouncing off a wall. This is shown in the upper diagram of the figure. As one quick look at the lower diagram of the figure suggests, something absolutely out of the ordinary takes place when the width of the barrier is thin enough, that is, not wide enough for the protruding portion of a matter wave to have died out completely within the barrier. The exponential tail has traversed the full width of the barrier and still has some "life" left as it reaches the other side. As the wave emerges to the open on the other side of the barrier, still retaining some amplitude, that is, some probability, it is capable of turning into another, much smaller in amplitude, matter wave. The wave, after having "pierced" an unscalable mountain, has simply gathered and resurrected itself on the other side of the mountain, and as long as the wave amplitude stands for the probability of finding the particle where the wave is, we will find the particle on the other side of a barrier just as if it had "tunneled" through. This is quantum tunneling. It is the whole story of this befuddling phenomenon.

An important point here is that a particle, be it an electron or a proton, has never physically burrowed through the barrier; it is energetically not possible. Instead, what has taken place is the strange interplay of the wave-particle duality and the subsequent uncertainties in which such a flow of mathematical probability allows it to happen. The particle is to be found wherever its probability does not vanish; if the probability can seep through, the particle can and will be found, even on the other side of an impenetrable barrier!

By its very nature, tunneling is sensitive to the width of a barrier, that is, the width of the energy profile of a barrier at the level of collision. Halving the width of a barrier, say, from 10 to 5 angstroms, or from 1 nanometer to $1/2$ a nanometer, increases the chance for tunneling by as much as a factor of about 10,000, in some cases by hundreds of thousands. In most cases, energy profiles of physical barriers are tapered up as shown in FIG. 6-10. This means that even a slight increase in the energy of an incoming particle can greatly enhance the chance for tunneling to occur. This secret of quantum tunneling lends itself to an extremely sensitive mechanism of controlling whether tunneling can take place or not. As you will see in the following two chapters, this property of tunneling plays the most crucial role in the technological application of this bizarre quantum phenomenon.

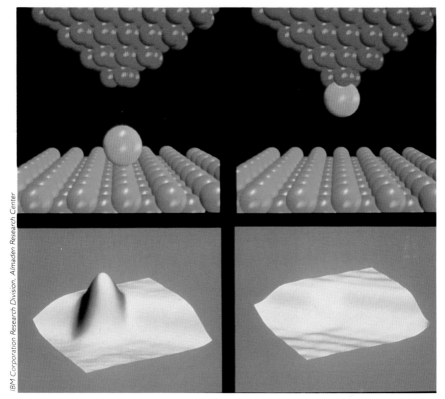

Computer-generated diagrams (top) represent the scanning tunneling micro-scope (STM) tip (red), a single-crystal nickel surface (blue), and a xenon atom (pink). At left, the xenon atom rests on the surface. An electrical pulse attracts the xenon atom across the gap to the tip (right). The actual STM images (bottom, magnified about 18 million times) show the presence (left) and absence (right) of the xenon atom on the nickel surface.

Scientists at IBM's Almaden Research Center created the world's smallest logo (660 billionths of an inch in length) by moving xenon atoms across a single-crystal nickel surface. IBM Corporation Research Division, Almaden Research Center

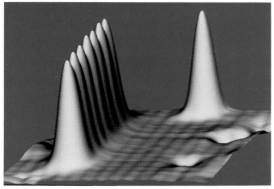

This row of xenon atoms, the first such manmade cluster, was assembled atom by atom by scientists demonstrating the ability of the STM to move and position atoms with great accuracy. The STM image is magnified about 13 million times.

IBM Corporation Research Division, Almaden Research Center

Scientists hope to use this xenon-atom "beaker," assembled on a nickel surface by an STM, as a type of corral in which they can form molecules.

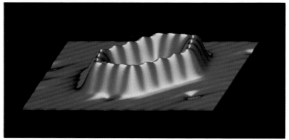

IBM Corporation Research Division, Almaden Research Center

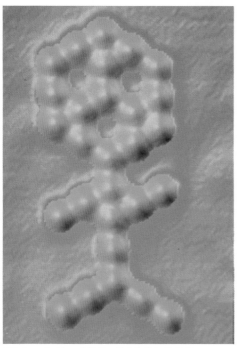

"Molecular Man" was created with an STM by assembling 28 carbon monoxide molecules on a single-crystal platinum surface. At 50 angstroms head-to-toe and 25 angstroms hand-to-hand, more than 20,000 figures placed side by side would be needed to span a single human hair.

IBM Corporation Research Division, Almaden Research Center

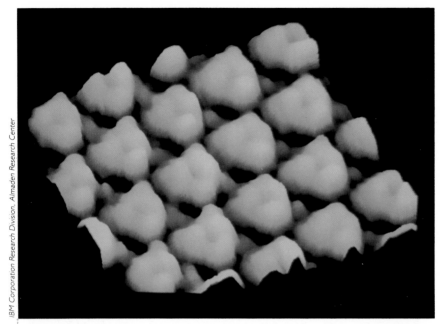

Each of the ring-shaped clusters in this STM image is a single benzene molecule. STM technology has provided visual reconfirmation that the atoms of a benzene molecule are arranged in a ring, just as chemist August Kekule envisioned in a dream in 1865.

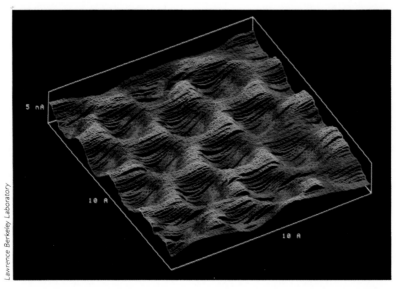

This STM image shows a layer of graphitized carbon atoms on a platinum surface. The surface shown measures 10 angstroms on each side and 5 nanoangstroms high.

Intel's i486 microprocessor, which packs 1.2 million transistors into an area approximately $1/2$-inch square, represents state-of-the-art microchip technology.

7

❋

The very small and the very cold

FOR MORE THAN SIX DECADES since its initial revelation as one of the first mathematical consequences of the wave nature of matter, quantum tunneling has remained as arcane as its physical origin. Not that it didn't have some dramatic moments of its own, but significant developments in the application of the tunneling effect were few and far between. All this changed dramatically when, beginning in the early 1980s, it was realized that the tunneling phenomena provided a unique mechanism to elevate some of the cutting-edge technologies to an unprecedented level of sophistication—such technologies as the synthesis of new materials and the microminiaturization of semiconductor devices. Suddenly one of the most abstruse aspects of quantum mechanics opened up possibilities to see and manipulate matter at the level of individual atoms.

In this sense the history of quantum tunneling parallels another little-known aspect of the quantum world, superconductivity. Since their discovery in 1911, superconductors have remained in the backwaters of technological advances. The field exploded in 1986 when a surprise discovery shook the world: A totally new class of materials—ceramic compounds, which are rather poor conductors of electricity under normal conditions—was discovered not only to superconduct but also to do so at temperatures much higher than anyone had suspected possible. This new breed of materials, popularly referred to as high-temperature superconductors (HTS), has since touched off an intense and competitive pursuits for technological and commercial break-throughs.

The confirmation of quantum tunneling as a physical process came swiftly enough, within two years of the work of Schrödinger. It wasn't the case that the effect was tested on some new experiment specifically devised for that purpose; instead, tunneling turned out to be the mechanism to solve a long-standing puzzle in naturally occurring radioactivity. Ever since radioactive materials were first discovered in the closing years of the nineteenth century, the question of what this radioactivity was all about had been shrouded in deep mysteries. In 1928, some three decades after radioactivity was discovered, George Gamow, perhaps better known for his initial proposition of the Big Bang theory, boldly interpreted one such radioactivity—the so-called alpha emission—as a manifestation of the tunneling phenomenon. In so doing, he was able to deduce a quantitative rule for this process, which remains the only theory for it to this date. It was not only a dramatic confirmation of the tunneling effect, but also one of the early triumphs of quantum physics.

Another major development occurred 30 years later, this time in the ultra-cold world of superconductivity. In 1962 Brian Josephson discovered an arrangement in which a superconducting electric current, that is, the motion of electrons through matter without running into any resistance from the atoms and molecules that make up solid matter, can tunnel through an insulating barrier interposed between two superconducting materials. A suitably constructed arrangement can thus turn on and off an electric current—a minute amount of a superconducting current—by allowing or disrupting the tunneling across the barrier. The device, called a Josephson junction, draws its significance from the fact that it can perform this switching function at speeds much faster than any available in microelectronics today. In this chapter, I discuss these two pre-1980 applications of quantum tunneling, one in the world of the very small (radioactivity of atomic nuclei) and the other in the world of the very cold (the switching capability—and its promises—of the Josephson junction).

The puzzle of lifetimes

The interpretation by Gamow in 1928 of a class of radioactivity as a manifestation of the tunneling effect was remarkable in more ways than one; not only has it solved a three-decade-old puzzle, but in doing so it demonstrated that the equation of Schrödinger, originally intended for solving the electron structure of atoms, could also be applied to goings-on inside an atomic nucleus at a time when little was understood about its structure.

Quantum mechanics had been born, after all, to help scientists understand the structure of atoms. In the scheme of things inside an atom, for all practical purposes, a nucleus was a sizeless mass point serving as the center of force for the "planetary" electrons orbiting around it. On the average, a nucleus is some 60,000 to 100,000 times smaller than an atom. By successfully applying the quantum effect to a nuclear process, Gamow has stretched, in one bold stroke, the limit of validity of quantum mechanics far beyond what was originally thought—down to the world inside atomic nuclei.

At the time, nuclear physics as we know it today was in the early stages of development. That a nucleus contained protons, and hence carried electric charges, and that a totally new force—a strong binding force never encountered before—was operative inside it had been known, but not much else. The existence of another constituent, the neutron, was discovered four years later, in 1932, by James Chadwick.

The atomic nucleus, as we have learned since, is a tightly held aggregate of protons and neutrons ranging from the smallest, the nucleus of a hydrogen atom (the proton), to the largest of the naturally occurring elements, the nucleus of a uranium atom with 92 protons and 146 neutrons. As FIG. 1-3 in the first chapter showed the characteristic scale for the nuclear sizes is a unit called a fermi, which is one millionth of a nanometer, or 10^{-15} meters. The fermi is 100,000 times smaller than the length of one angstrom, 10^{-10} meters. Most atoms range between about 1 and about 4 angstroms in size. The sizes of atomic nuclei, on the other hand, range from about 1 fermi for the radius of either a proton or a neutron up to about 7 fermis for the uranium nucleus. This spread in nuclear sizes tells you right away how tightly packed objects the nuclei are: All 238 of them, the 92 protons and 146 neutrons of a uranium nucleus, are stuffed into a volume with a radius of only seven protons. What you have is a soccer ball stuffed with 238 pingpong balls!

As is almost always the case, we illustrate these incredibly tiny atomic nuclei by drawing "pictures" like the ones shown in FIG. 7-1: a proton and a neutron forming a deutron, which is the nucleus of deuterium, also known as heavy hydrogen; a particle, originally named an alpha particle and later identified in 1909 as the nucleus of a helium atom, made of two protons and two neutrons; or a larger nucleus, which begins to look like a tightly bundled ball of grapes.

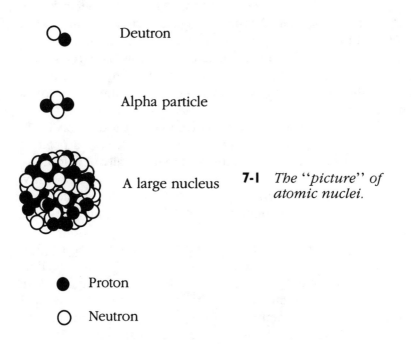

Deutron

Alpha particle

A large nucleus   **7-1** *The "picture" of atomic nuclei.*

● Proton

○ Neutron

These pictures are, of course, just that: a convenient visual aid. However much we would like to cling to the picture of particles as nice, round, hard balls, we know by now that the reality of quantum world denies it. We use them anyway for their sheer ludicrous convenience. Try to imagine a picture of an alpha particle as an assembly of four wiggling waves, two positively charged waves superimposed on two electrically neutral fluffy waves!

The size of a proton is determined in the same way as the size of an electron: by the farthest extension of its electric charge distribution as can be checked by experiment. It comes to about 2 fermis, or 2×10^{-15} meters, across, making a proton at least 1000 times larger than an electron. Often the size of a proton is given as having a radius of 1 fermi, the word "radius" implicitly implying a picture of a sphere. Quantum-mechanically speaking, it is the extension of a proton wave, the farthest reach of its probability distribution, that stretches about 2 fermis. The size of a neutron is about the same as that of a proton.

Whereas the mass of either a proton or a neutron is about 1900 times larger than that of an electron, when it comes to the comparison of electric charges, an electron is an equal match to a proton. The positive electric charges residing on a proton are exactly the same in magnitude as the negative charges on an electron. A neutron, of course, has no net electric charges; it is the neutral cousin to the proton. The best known current values for a proton are:

Size: about 1×10^{-15} meters in radius

Mass: 1.672×10^{-27} kilograms

Charge: $+(1.602189 \pm 0.0000046) \times 10^{-19}$ coulombs

The values for a neutron are:

Size: about 1×10^{-15} meters in radius

Mass 1.675×10^{-27} kilograms

Charge: Zero.

One important aspect of the family of atomic nuclei is that the number of its species is limited. For reasons stemming from an interplay between the forces that operate within the bounds of a nucleus, when the number of constituent protons and neutrons reaches a certain point, a built-in mechanism of natural instability sets in. Depending on who is doing the counting, the total number of elements in the entire universe is either 105, the last and the heaviest being the artificially produced and short-lived hahnium with 105 protons and 155 neutrons, or as high as 112, if you include the so-called superheavies. In any case a definite limit exists beyond which protons and neutrons simply cannot all hold onto each other to form and maintain a cohesive nucleus.

Most of the elements we are familiar with are both stable and occur naturally, but others found in nature are unstable—these are said to be naturally radioactive—while still others are neither stable nor found in nature: the artificially produced ones. Only 81 elements are naturally stable, and 11 elements are naturally radioactive and have very short lifetimes.

Instability begins to set in, as a general rule, just beyond the nucleus of a bismuth atom, which, with 83 protons and 126 neutrons, is the heaviest stable nucleus around. The nine elements beyond bismuth, which include polonium, radon, radium, thorium, and uranium, are all naturally radioactive. In a set of characteristic times, which differ from one nucleus to another over a wide range from a few microseconds to a few million years, these naturally radioactive nuclei shed their excess and contract down to more stable configurations. This is called natural radioactivity, the spontaneous emission of excess particles and energies. The heaviest of these, the uranium nucleus, is so unstable that a slight disturbance, such as a collision with a wandering neutron, can split it apart in pieces—the process called nuclear fission. Without the mechanism of this built-in instability, it would have been possible to have in nature a nucleus comprised of, say, 100,000 protons and 300,000 neutrons. We might have had an atom that we could weigh on a bathroom scale!

The discovery of natural radioactivity traces its origin to 1896, when Henri Becquerel observed a strange penetrating radiation emanating from a sample of uranium salt. Within two years Pierre and Marie Curie had identified two more radioactive elements, polonium and radium. All three shared the Nobel physics prize in 1903 for their discoveries, which marked the beginning of what was to become the nuclear physics of today. It soon became clear that this radiation was not of one kind but actually contained three separate varieties, distinguishable from each other by marked differences in their power of penetration through matter. While the least penetrating type could barely go through a sheet of paper, another component could penetrate through up to an $1/8$ inch of aluminum, and the most powerful could easily pass through 3 to 4 inches of lead. Tentatively named alpha, beta, and gamma rays in the ascending order of strength, after the first three letters of the Greek alphabet, this trio of nuclear radioactivity still retains those original designations to this day, long after their identities have been determined.

Gamma radiation, also known as gamma emission or decay, refers to the process in which highly energetic photons, the quanta of an electromagnetic radiation as discussed in chapter 3, are emitted by an atomic nucleus as it undergoes a sudden transition from higher to lower energy levels. The process is essentially the same by which atoms emit photons, but the photons emitted by nuclei have energies a few orders of magnitude greater. Atomic photons have energies in the range of visible light, ultraviolet, and some soft X-rays, while the nuclear photons correspond to the part of the spectrum that contains hard X-rays and gamma rays, the very region defined by gamma radiation. These nuclear photons can be harnessed, under the right conditions, to generate X-ray laser beams, which constitute the mainstay of the antiballistic missile defense project popularly known as the "Star Wars" program.

The radiation of intermediate strength, the beta radiation or beta decay, was the first among the three to be identified; by 1900 it was proved as a beam of electrons, themselves discovered only three years earlier. Under suitable conditions, a constituent neutron inside a nucleus undergoes a transmutation into a proton, and as it does the nucleus emits an electron, balancing out the electric charges of the process. An electron, however, does not reside within a nucleus; it is created—and ejected—at the moment of the transmutation of a neutron into a proton. Much later, that is, in the 1930s, it was also observed that a constituent proton could match the neutron's trick by pulling the reverse: A proton transmutes itself into a neutron and this time a positively charged "electron" is created and ejected. This is the positron, the antimatter of an electron. This second process is the standard source of positrons for a cutting-edge medical imaging technology known as positron emission tomography, the PET scan for short.

The least-penetrating of the three types of radiation turned out to be a beam of doubly charged objects, named the alpha particles, ejected from within a radioactive nucleus. This disintegration process is still referred to by its original name, alpha emission or alpha decay. Not until 1909 did Ernst Rutherford conclusively demonstrate the identity of an alpha particle as a doubly ionized helium atom. The alpha particle played a preeminent role in the development of the atomic picture of nature. Through the so-called alpha scattering experiment, in which a beam of alpha particles was shot into a thin metallic foil and their deflection patterns pinpointed a hard core at the center of an atom, Rutherford in 1912 discovered the existence of the atomic nucleus itself. Following the discovery of neutrons in 1932, it became apparent that an alpha particle was a helium nucleus that comprised two protons and two neutrons. It is a very tightly bound nucleus that

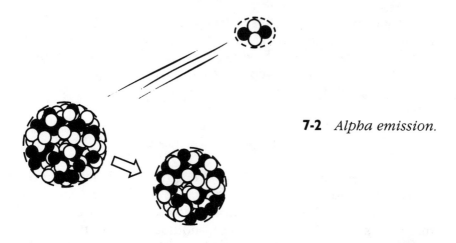

7-2 *Alpha emission.*

larger radioactive nuclei prefer to eject to reduce their degree of instability.

An emission of an alpha particle clearly represents a break-up of the original nucleus, not a violent disintegration into several flying fragments, but a break-up nevertheless. And whatever mechanism is responsible for this action, it originates from within a nucleus, something 100,000 times smaller than an atom. The quantum rules of probability should certainly prevail in such a small microcosm and indeed they do. Even a miniscule sample contains millions of radioactive nuclei. They do not emit alpha particles in unison as if on cue, nor, for that matter, on any fixed determinable schedule. Each nucleus does its own thing when it is good and ready in a probabilistic way, and it is not possible for us to either calculate or know exactly when a particular nucleus will emit an alpha particle. The principle of probability applies not only to the question of where—the position of the particle— but also to the question of when a particular process might take place over a period of time. We can measure the rate at which some approximate number of nuclei would have emitted alpha particles over a given period of time on a statistical basis.

This rate of transition, the rate with which a certain number of nuclei in a sample will have emitted the alphas over a given period of time, is expressed in terms of the so-called half-life. A half-life of a radioactive nucleus refers to the time period, an experimentally determined quantity, during which half of the original amount of radioactive nuclei will have transmuted themselves by emitting radiation. We can measure the half-life for alpha

Table 7-1
*A radioactive material
with the half-life of 10 years.*

Year	How much left in pounds
1950	16
1960	8
1970	4
1980	2
1990	1

emission, the half-life for beta emission, and so on. Suppose we have nuclei whose half-life for alpha emission is 10 years. As shown in TABLE 7-1, if we had started with 16 pounds of this sample in 1950, one pound would have not yet emitted alpha particles by 1990.

The puzzle of alpha emission, which perplexed scientists for three decades, is the puzzle of lifetimes; that is, the vast differences in half-lives that exist among a group of otherwise quite similar nuclei. Some typical half-lives of alpha emission, as determined by measurements, are shown in TABLE 7-2. The vast disparity among them is obvious: It stretches from a few microseconds for polonium-212 to the near-stable half-life of a few billion years for uranium-238. Long half-lives have been calibrated on the basis of observations over a shorter and more manageable period of time. Of course, had it not been for some of the longer half-lives, we wouldn't have any radioactivity left today to talk about; all naturally radioactive elements would have disappeared in the early days of the universe!

Table 7-2 *Half-lives of some alpha-radioactive nuclei.*

Nucleus	Number of protons	Number of neutrons	Half-life
Uranium-238	92	146	4.5 billion years
Uranium-234	92	142	250,000 years
Thorium-230	90	140	80,000 years
Radium-226	88	138	1600 years
Radon-222	86	136	3.8 days
Polonium-218	84	134	3 minutes
Polonium-214	84	130	1.6×10^{-4} seconds
Polonium-212	84	128	3×10^{-7} seconds

The naturally radioactive nuclei all share a similar structure. They are relatively large, containing about 210 to 240 constituents, about 80 to 90 protons and about 130 to 150 neutrons. For the alpha emission, the similarity among them cuts deeper: The energy of the alpha particle, that is, the speed with which they are ejected, is similar among nuclei, deviating no more than 50% from either side of the average. The fact that identical particles are ejected with roughly the same energy from a group of similar nuclei clearly indicates that the same mechanism must be operative inside these nuclei. Yet the same mechanism that prompts a nucleus to spit out an alpha particle in a matter of a few microseconds can hold it off for another nucleus for as long as a few billion years! This remained a puzzle until quantum tunneling came to the rescue.

An alpha tunnels through

To see how an emission of an alpha particle turns out to be an act of quantum tunneling, that is, an alpha particle tunneling through an energy barrier at the surface of a radioactive nucleus, we must now seek to understand what the inside of a nucleus is like—its constituents, structure, and underlying dynamics. The structure of a nucleus differs from that of an atom in several ways. As we have already mentioned, a nucleus is a very tiny thing, some hundred thousand times smaller than atoms, and into this miniscule volume of a space are packed protons and neutrons, squeezed in tightly—a sort of a lumpy ball formed by squeezing between your palms scores of little balls of toy putty.

Each nucleon, as protons and neutrons are collectively called, is held in place by neighboring nucleons—each little putty ball sticks to adjacent putty balls and together they hold themselves as a nucleus. Unlike atoms, each of which has a fixed center known as the nucleus, no such center of force exists inside a nucleus. The constituent nucleons are all equal partners, each sticking to its immediate neighbors. This structure of a nucleus leads us to its most important and dramatic secret, the revelation that within the boundaries of a nucleus exists a totally new kind of a force never before seen or felt. Packed inside nuclei ever since the early moments of the universe, this force stayed hidden from our view for all this time. It holds together not only a pair of neutrons or a neutron and proton, but also a pair of protons, evidently overcoming the strong repulsive electric force that acts between them.

The strength of the electric force with which one proton inside a nucleus repels another is no laughing matter. The force between electric charges obeys a precise mathematical law, the famous inverse square law. It is proportional to the product of charges and inversely proportional, to the square of the distance between them. The force thus decreases and gradually fades as the distance gets larger, but increases as the gap between charges narrows, rising to an enormous magnitude in the scale of atoms and nuclei. Between a pair of protons separated by, say 2 fermis, that is, 2×10^{-15} meters, the repulsive force comes to a little more than 10 pounds, and that is a lot of push for a little proton. An alpha particle, being doubly charged, would feel a repelling force of 40 pounds from two nearby protons joining forces to push it out. A force of 40 pounds is not something trivial even for us, but for an alpha particle with a mass of 4×10^{-24} grams, it is nothing short of a cosmic blast!

The force that holds nucleons together in a nucleus should be at least as strongly binding as the electric force is repelling. A real surprise about this new force—once called the nuclear force, then the strong force, and now the strong nuclear force—is that its strength lives up to its name. It is roughly 100 times stronger than the corresponding electric force. A pair of protons—or a pair of neutrons, or a neutron and a proton—about 2 fermis apart experience a binding nuclear force of about 1000 pounds, which absolutely overpowers the electric force. This brute strength, the enormous power of the strong nuclear force, is the source of the colossal amount of energy packed inside nuclei. Its release—controlled or otherwise—is what we call nuclear energy. With such a strong attractive force, it would be possible to form a nucleus with, say, one million nucleons. If a lumpy ball can be formed by pressing together about 40 or so small putty balls between our palms, why not add more putty balls to turn it into a larger ball, the size of a snowman or a medicine ball? Here we see another wonder of nature—a delicate balancing act—that keeps the brute power in check by severely limiting the range of its effectiveness. Had it not been for this severe limit imposed on the strong nuclear force, the universe would be quite different from what it is.

The range of a force means what it says—the range of distances over which the force is effective. A definite range for a force implies a maximum distance beyond which the force would abruptly disappear. It is a simple concept that is, however, unfamiliar. Two forces of nature with which we are familiar, the gravitational and electromagnetic forces, are not hampered by any fixed range, that is, there is no fixed point beyond which they drop to zero. Both these forces obey the inverse square law. While the strengths of the forces decrease rapidly with increasing distances, as they should, the decrease is very gradual with a long tail, as shown in FIG. 7-3. We call these forces long-range forces, although they actually have an infinite range.

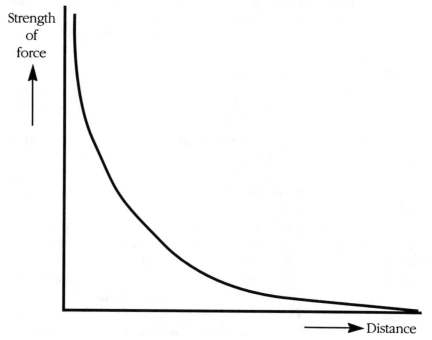

7-3 *The long-rangedness of the inverse square law force.*

However, the strong nuclear force is kept in check by an extremely short range beyond which the force simply does not exist. Its range is so short that referring to it as a short-range force, as is customarily the case, is something of an understatement. The maximum range is determined to be no greater than about 2 fermis, that is 2×10^{-15} meters. While a pair of nucleons whose centers are 2 fermis apart are bound to each other by the fiercely attractive force, a third nucleon standing around, say 4 fermis away wouldn't even be aware of the pair. Lacking an example of such a strong but short-ranged force in our daily experience, let's consider as an analogy a powerful tornado cutting a narrow, meandering path and unleashing an awesome power of destruction. Let's say that the path of destruction is about one block wide. Everything within its path is a total loss, yet a house just a block away from its path escapes unscathed. It is a powerful force in action, but its range, its lateral range, is only one block short.

The constrasts between the two forces, the electric repulsion and the strong nuclear attraction, are almost as opposite as day and night, and the combination of the two inevitably leads to the natural built-in instability for larger nuclei. The strong nuclear force exerts an enormously powerful binding between nucleons, protons, and neutrons alike, but ceases to exist just beyond an extremely short distance. The electric force, not hampered by any notion of a limited range, can bring to bear on a proton or an alpha particle inside a nucleus the total combined force of repulsion from every other proton. One or two big pulls against a combination of pushes sooner or later leads to an unstable situation. This is the basic principle of nuclear instability.

Suppose we have a nucleus with 81 protons and 124 neutrons. Any one of the protons feels the collective repulsive force from 80 other protons, all trying to push it out. Suppose the proton in question finds only one other nucleon, its nearest neighbor within the nucleus, within the range of the strong nuclear force. If we subtract the two numbers, a repulsion by a factor of 80 from an attraction by a factor of 100, we realize that the proton in question is hanging in there precariously. To exaggerate a little, a gentle sneeze might just pry that proton loose! An alpha particle that is doubly charged (two protons) has to contend with a much stronger force of electric repulsion. This instability arising out of the intricate interplay between the electric and the strong nuclear forces is what limits the size of naturally occurring nuclei; we can only place so many protons in a nucleus before it begins to spit out either protons or alpha particles and, at a higher level of instability, even split itself into many fragments. The process is called spontaneous nuclear fission. Contrary to a widely held view that "fissionability" is something invented by scientists, it is something that was put into—and kept secret inside of—nuclei in the very beginning of the universe some 15 billion years ago.

Now we come to the crux of the matter: how the puzzle of the alphas—the vast disparity in the half-lives of alpha emissions by various radioactive nuclei—was solved by the phenomenon of quantum tunneling. Once you understand the mechanics of the tunneling effect, the answer to this puzzle appears when we examine the nature of the two counteracting forces near the surface of a nucleus. An alpha particle represents a particularly tightly bound configuration of four nucleons, two protons and two neutrons, that can be considered to be a doubly charged single object. Just as I described in the last chapter the force of resistance a golf ball would encounter coming upon a patch of grass or a sand pit in terms of the energy profile and the barrier that it represents, let's see what shape barrier an alpha particle would encounter at or near the surface of a nucleus.

Figure 6-8 in the previous chapter, the energy profile of a solid wall, shows that the barrier a handball bounces off has a tall and thin shape. For a ball with a nominal amount of energy, it is an impenetrable barrier, representing a strong repulsive force. Also, it is "short-ranged" in that the wall is narrow. The strong nuclear force has a short range, but it is a force of attraction instead. Its energy profile is represented, not by a narrow and high tower, but instead by a narrow and deep well. Once you have fallen into a deep well, you have little chance of getting out. You are trapped. You are bound by a strong confining force. Taking the enormous strength and very short range into account, a typical energy profile for the strong nuclear force looks like a narrow and deep well, as shown in diagram (a) of FIG. 7-4.

Away from the surface of the nucleus, there is only one force to reckon with: the electric repulsion that works against any charged particle approaching the nucleus. Its energy profile follows the general shape of the force as shown in FIG. 7-3, except that it would stop at the outside surface. This is shown in the diagram (b) of FIG. 7-4. From this point on, things become more or less self-evident. The energy profile of the barrier for a charged particle such as an alpha particle in and out of a nucleus is the combination of the two, (a) and (b), as shown in the diagram (c) of FIG. 7-4. This barrier confines alpha particles within the size of a nucleus. This barrier, with its sharply tapered shape and gradual tail, provides an almost ideal setting for quantum tunneling to do its wave-probability tunneling-through thing! As an alpha particle bangs against the barrier from inside a nucleus, it hits the wall at different heights, that is, at different levels of energy, and encounters different widths of the barrier, some narrow enough for tun-

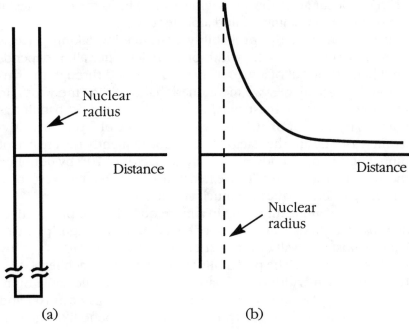

(a)

Energy profile of
the strong nuclear force

(b)

Energy profile of
the electric force

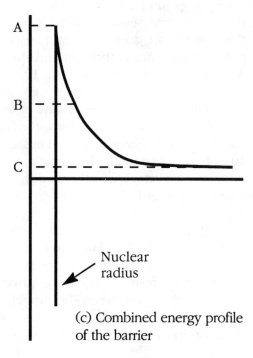

7-4 *The barrier for the*
alpha emission.

(c) Combined energy profile
of the barrier

neling to occur and others too wide for it. This is the dynamics of the quantum tunneling behind alpha emission.

By its nature, the probability of tunneling taking place is extremely sensitive to the width of a barrier. After all, it depends on whether the tail of a matter wave can extend through the barrier. In the last chapter we mentioned that doubling the width, in the atomic scale of angstroms, can cut down the probability as much as by a factor of 10^5. In the nuclear scale of fermis, sensitivity increases by another factor of 10^5. Consequently the chance of tunneling can be cut down as much as a factor of 10^{10} by doubling the width of a barrier. The extension of the matter wave of a nuclear particle is, after all, much smaller.

Let's put alpha tunneling in a more quantititiative perspective. Suppose we choose three levels of energy at which alpha particles bang against the wall of the barrier, as indicated in FIG. 7-4 (c) by levels A, B, and C. The probability of tunneling at each level, that is, against each width of the barrier, is a readily calculable number: Let's say they are one chance in 10^{14}, one chance in 10^{22} and one chance in 10^{38}, respectively. That doesn't sound like much of a chance until we figure how many times per second the alpha particles can be estimated to bang against the wall. A simple estimate places this number to be about 10^{21} times of collision against the wall in one second. Multiplying the chance of tunneling by this number, we get a good estimate of the actual number of alpha particles that tunnel through and show up on the other side the barrier: 10^{21} times one chance in 10^{14}, that is, 10^{21} number of tries times the probability of tunneling, yields no less than 10^7 alphas tunneling through in one second at level A. This translates to an average time of 10^{-7} seconds for the first alpha to tunnel through. Likewise at the level B, 10^{21} times one chance in 10^{22} gives us 10^{-1}, that is, one tenth of a particle, tunneling through in one second, or an average of 10 seconds for the first alpha to tunnel through at level B. Tunneling becomes very scarce at the level C: 10^{21} times one chance in 10^{38} equals 10^{-17} particles tunneling through in one second, in other words, it would take 10^{17} seconds for a single alpha particle to manage to come through at the level of this energy. So we have 10^{-7} seconds, 10 seconds, and 10^{17} seconds, respectively, at the three levels. This is a difference in factor of 10^{24} between the tunneling probabilities at levels A and C. A great disparity in lifetimes is indeed possible for a small variation in the energies of the emitted alpha particles involved.

The notation in the powers of 10 can easily fool our perception, and it is a little difficult to grasp just how long 10^{24} seconds is until we realize that one year is only 3×10^7 seconds. The half-life of alpha emission for uranium-238 is $4^{1}/_{2}$ billion years, a very long time when expressed in terms of a "mere" 1.4×10^{17} seconds. The difference between this and the half-life of polonium-212, which is 3×10^{-7} seconds, corresponds to the difference by a factor of 10^{24} seconds. You can see how easily the tunneling phenomena can handle such a vast difference. This is how scientists solved the puzzle of the lifetimes of radioactivity called alpha emission. The success was not only a dramatic confirmation of the quantum tunneling effect, but also a smashing triumph for the physical principle underlying it, the wave nature of physical entities, routinely called particles, at the level of the quantum world.

The triple-super switch

We now come to the second of the two pre-1980 applications of the tunneling phenomenon: a little-known device called the Josephson junction switch. The reason for its obscurity becomes apparent when we realize that the device operates on the combined principles of tunneling and superconductivity, one subject just as esoteric as the other. To describe it in a single sentence, we need to link together several "super" words: It is essentially a supercold, superfast, and superconducting digital electronic switch, a triple-super switch indeed—a quantum-tunneling triple-super switch at that.

First discovered in 1962 by Brian Josephson, who was awarded a share of the 1973 Nobel prize in physics for his work, the device has an enviable capability of being able to switch an electric current at an incredibly fast speed, up to a few trillion times a second. This is hundreds or even thousands of times faster than some of the fastest switching devices used in microelectronics today. Despite this great potential, partly due to impracticalities involved in dealing with superconducting materials in a supercold environment, the research and development effort for this device has sputtered along for some 25 years now without its technology ever reaching mass production and commercialization. All this is about to change. The race is now on in advanced research laboratories around the world to turn this experimental device into a full-fledged workhorse commodity for a new age of computers. Let's find out what this mysterious device that combines the quantum phenomena of tunneling and superconductivity is all about.

The conductivity I am speaking of here refers to the electric conductivity of a material, the capacity in which it can serve as a channel for the passage of an electric current. It turns out that this property is not evenly distributed among all substances. With some exceptions, notably the naturally occurring semiconductors, most elements belong to one extreme or another: Either they are excellent conductors of electricity or they are very good insulators. Of the latter, the most familiar ones include such well-known insulators as glass, quartz, paper, wood, amber, and porcelain. At the other end of the scale, all good conductors are metals: silver, copper, aluminum, iron, nickel, and alloys such as brass and steel, with silver being the best conductor of all. This fact notwithstanding, the use of silver wiring is not widely practiced for an obvious reason.

An electric current is a flow of electric charges just as a river current is a flow of water molecules. As we plug in an appliance or connect it to a battery, the collective motion of a whole bunch of electrons constitutes an electric current. One of the things that characterizes a good conductor, therefore, is the abundance of loose electrons, also called the free electrons. These are the electrons on the outer periphery of an atom that are pulled every which way by other nearby atoms in the densely packed innard of a solid, and consequently are quite loose. Typically, a good conductor can boast an average density of about 10^{22} free electrons per cubic centimeter, that is, 10 billion trillion of them in a cubic chunk of, say, copper with sides one centimeter long. Let's make the cube a tad smaller and imagine a copper cube with sides 1/8 by 1/8 by 1/8 inch; you would have in it about 3×10^{20} free electrons, that is, 300 billion billion free electrons, ready to move. A good insulator, meanwhile has no more than 100 free electrons per cubic centimeter or, for practical purpose, zero free electrons. In the scale of things we are talking about, in which a hand-held pocket calculator passes about two trillion electrons through its circuity in one millisecond, a mere hundred electrons is simply nothing.

Another factor that has a bearing on conductivity is the density of obstacles that the free electrons encounter as they pass through a maze of endless lattices of atoms and molecules, which is what a solid is. Down a straight and smooth conduit, a water current will flow freely and fast, without any hindrances or impeding vortices. But coming upon a narrow and winding channel filled with jutting rocks, the flow would slow down to a white water, some water splashing out and scattering into a mist. The flow of electrons through the interior of a solid is anything but a smooth and orderly march. Instead, they collide constantly with atoms, molecules, and other electrons. Bouncing around from these endless collisions, each electron moves every which way, sideways and back up the stream, but collectively the electrons manage to drift in one direction, forming what we call an electric current. The sum of these collisions not only slows the flow, in doing so they rob energy from the current in a process we call electric resistance. Some material is less resistive than others, but all materials as a general rule have a resistance just as surely as they have structures. The better a conductor is, the less its resistance and the less energy and power are lost by an electric current flowing through it.

The energy continually drained away from a flow of current through a resistive medium turns into heat and is quickly dissipated. This unavoidable conversion of electric energy into heat energy is the source of either a great benefit or a wasteful loss, depending on its use. As the source of electric heating, the resistance is an indispensable property of matter that not only provides us with all manner of creature comforts, but also, at a much higher level of resistance, electric lighting, in the case of incandescent light bulbs. But in all nonthermal applications of electric energy—from a hand-held electric toothbrush to large industrial machines such as power generators, power motors, transformers, and electromagnets—electric resistance is the main culprit responsible for reduced efficiency resulting in wasted energy. A substantial portion of the power disappears as the electric current is transmitted over high-voltage power lines, heating the air surrounding the cable. Even such mundane tools as the electric weed cutter or leaf blower can provide you with a "live" demonstration of resistance at work. Manufacturers of these appliances do not usually recommend their operation beyond 100 feet of extension cord and for a good reason. At about 200 feet of extension cord, you will notice a marked drop in power when your weed cutter just can't cut anymore.

It would do wonders for efficient use of electricity if we

could find a material which, under any condition, even temporarily possesses a dream capability of being able to pass current with no resistance. It would be an electric counterpart of a perpetual motion machine, currents flowing forever without ever losing power, sort of like a hockey puck sliding effortlessly over a frictionless surface forever without slowing down. No resistance means that trillions of free electrons could move through the endless maze of lattices without a single collision! Astoundingly, just such an impossible dream turns out to be a reality, albeit in an extreme environment. At extremely cold temperatures, say −400 degrees Fahrenheit, certain materials abruptly lose all resistance in what is called superconductivity. It is more than just *super* conductivity; it is the state of conduction with zero resistance. If you could lay your hands on a superconducting extension cord, you could blow leaves in New Hampshire while still plugged into an outlet in Albuquerque!

Over the past several years we have heard a lot about superconductivity as its commercial applications seemed more realizable and the accompanying media hype rose. Suddenly everyone was talking about it, or so it seemed. It all began in the fall of 1986, when a bombshell of a discovery sent shock waves around the globe. Lifting the secrecy from their work, K.A. Muller and J.G. Bednorz announced to the world their dramatic discovery of a previously untested material, a ceramic oxide compound that began to show superconductivity at a temperature much higher than any previously recorded with any material.

It was the first time in at least 25 years, since the early 1960s, that anyone had been able to raise the temperature at which the onset of superconductivity was observed. The highest temperature up to that point was −418 degrees Fahrenheit, observed with a special niobium-germanium compound. The significance of this 1986 discovery lies in the fact that it showed, first, that the old temperature barrier could be broken with the help of ceramic superconductors and, second, that the great promises of superconductivity can perhaps be realized in a more manageable and economically practical environment. The use of the words "warm," "high," and "cold," in the world of superconductors is relative. A temperature of −418 degrees is 36 degrees warmer than −454 degrees Fahrenheit, which is the temperature of interstellar space.

The news of the ceramic oxide superconductors immediately

touched off a flurry of activities worldwide and ushered in a "gold rush" period of two years, 1987 and 1988, in which the pursuit of newer materials and yet higher threshold temperatures became a sustained frenzy, one dramatic event following another at a breathless pace. Within a year came the discovery of a slightly different copper oxide compound that showed the onset of superconductivity at a hitherto unheard-of temperature of −290 degrees Fahrenheit. This impressive record, however, did not last long. In January 1988, a version of a thallium oxide compound was successfully tested to begin to superconduct at −234 degrees Fahrenheit, which remains to this date the highest confirmed threshold temperature for the onset of superconductivity.

Hardly a month went by without someone somewhere claiming to have discovered yet another higher threshold temperature, only to remain unconfirmed. This eventually led to rumors that room-temperature superconductors would require no refrigeration whatsoever. Things somewhat quieted down by 1990, the arena of action having moved out of the laboratories and into the patent office and courts for disputes and litigations of claims and counterclaims. Many of these are pending to the time of this writing. With all the hoopla and hype, you would have thought that superconductivity was discovered for the first time in the fall of 1986!

The fact is that the discovery of superconductivity traces its origin back to 1911, when Heike Kamerlingh-Onnes noticed an abrupt and complete disappearance of electric resistance in a sample of mercury as it was chilled, immersed in a liquid helium, to the temperature of −451 degrees Fahrenheit. This is just 8 degrees Fahrenheit above absolute zero. The same effect was soon observed with several other materials: lanthanum, thallium, and aluminum at −451, −455, and −457 degrees Fahrenheit, respectively.

Superconductivity is truly a marriage of two extremes, an extreme of zero resistance attainable only at another extreme of supercold temperatures. It is also exclusive in that it is not a universal property of all matter. In fact, most materials never reach that stage at any temperature, no matter how low. It is a property that only a handful of select materials display when cooled below a certain threshold, called the critical temperature, each having its own unique threshold. Superconductivity is not so much a property as it is a new phase of a matter—a fourth state, as in gaseous, liquid, solid, and superconducting states. Just as a liquid freezes into a solid at its freezing temperature, some solids "superfreeze" into superconductors at their critical temperatures. The critical temperatures are so low, moreover, that the state of superconductivity, once attained by extensive cooling, is a fragile one. With only a slight warming, the material reverts to a plain solid.

The fragility of superconductivity depends not only on the variation in temperature, but also on changes in two other external factors. It is extremely sensitive to both the magnitude of the superconducting current itself and the strength of the magnetic field surrounding a superconductor. Taking full advantage of no resistance, if we try to pass too much current, at some point superconductivity suddenly disappears altogether even while the material is kept at a temperature well below its critical temperature. This threshold in the amount of the superconducting current is appropriately called the critical current. Critical indeed! How much good is a superconductor if it cannot carry sufficient current to carry a workload that justifies the expensive refrigeration requirement? In fact, the rather disappointingly low level of critical currents for the recently discovered higher-threshold ceramic superconductors continues to be the single most difficult impediment to their full commercial development.

The third fragile aspect of superconductivity has to do with the fact that a magnetic field of a suitable strength can also suddenly kill it off. Suppose we introduce an externally generated magnetic field to a superconducting setup doing its thing—at below its critical temperature and carrying less than its critical current—and gradually increase its strength, say, by raising an electric current on an ordinary wire that is producing the magnetic field. As the strength of the magnetic field reaches some point, superconductivity suddenly vanishes; this is the critical field value of superconductivity. The three interlocking thresholds—critical temperature, critical current, and critical field— place tight constraints on superconductivity. As shown in FIG. 7-5, they combine to allow for a small window for maintaining it.

The sensitivity of superconductivity to a slight variation in a magnetic field, coupled with the relative ease with which we can control its strength, lends itself readily to a convenient and useful mechanism for a variety of applications. It can be used as a sensitive device to detect changes in a magnetic field or as a switching device to turn on and off a superconducting current. In the former

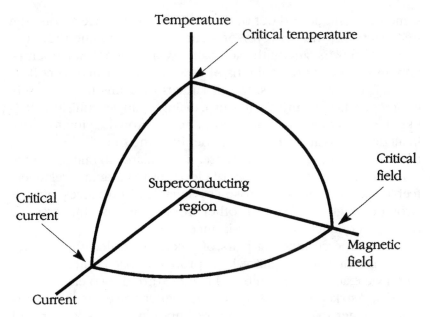

7-5 *The three thresholds of superconductivity.*

category, a sensitive superconducting magnetometer can be employed for exploring underground oil reserves or detecting submerged submarines. But its most promising application lies in the latter category as a switching mechanism for a superconducting current, a totally new kind of digital switch that can operate much faster than conventional silicon-based semiconductor switches. This is one of the two key ingredients that make up the Josephson junction switch, the other being the unique manner by which superconductivity enhances the chance for quantum tunneling.

Although we have gained a good deal of understanding over the years, we still do not have a complete theory of superconductivity that explains the behavior of the old (pre-1986) and new ceramic oxide superconductors. The first, and so far the only, theoretical understanding of them came in 1957, some four and a half decades after the initial uncovering, when John Bardeen, Leon Cooper, and Robert Schrieffer at Bell Laboratories advanced what at the time was considered a comprehensive theory known as the BCS theory, for the initials of their last names. Its validity was restricted, however, to a range of temperatures that is extremely low even by the standards of the superconducting world.

According to this theory, at the onset of superconductivity, that is, just as a material is chilled down to its critical temperature, something unique and out of the ordinary takes place among the free electrons, the conveyor of a current. As if on cue they suddenly form pairs, and with each pair—two electrons in tandem—now moving as a unit, trillions of these newly formed pairs flow through the frozen landscape of lattices of atoms in a precisely choreographed manner without causing a single collision. It is like the mass of an uncontrolled mingling crowd trying to move in all directions at once when suddenly, at the sound of a whistle, they fall into straight lines two abreast and march in military precision! The link that holds a pair of electrons in tandem is relatively feeble and can readily be broken by suitable disturbances, but it is strong enough to steer each other from collisions, thus creating the superconducting, zero-resistance state.

The superconducting pairs of electrons, called the Cooper pairs, harbor another marvelous property that comes to light when we examine the combined matter wave of two electrons in tandem. As sketched in FIG. 7-6, the matter wave for the pair is much further extended and, more importantly, has longer and more gradually sloping tails than that for an individual electron. To the unintitiated this might not appear to be a big deal; that one wave is larger than the other hardly seems to be anything out of the ordinary. To those who understand the workings of the quantum tunneling phenomenon, however, this is a hidden treasure with an important implication: A superconducting pair with a longer wave tail can penetrate or tunnel through a thick barrier where no individual electron can. The superconducting current, by virtue of being the flow of pairs of electrons, enjoys an en-

A single electron wave

A matter wave for a pair of electrons

7-6 *A pair of superconducting electrons versus a single electron.*

hanced probability of the tunneling effect compared to an ordinary current of single electrons. This is the other key ingredient that makes the Josephson junction switch possible.

The simplified schematic of a Josephson junction switch shown in FIG. 7-7 illustrates the two key mechanisms that make it work: an enhanced tunneling of a superconducting current through an insulating barrier and a control current nearby to create a magnetic field that would shut off the superconductivity—a basic on-and-off proposition. The thin insulating material sandwiched between two disjointed superconducting wires is only a few nanometers thick, that is, several billionths of a meter, but it is thick enough to completely block any tunneling by individual unpaired electrons. No flow of individual electrons can cross the barrier, tunneling or otherwise. As the material is chilled below its critical temperature, electrons coalesce into pairs and, by virtue of their extended waves, begin to tunnel across the barrier, giving rise to the flow of a current—a minute amount of tunneling current, but a current nevertheless. And this current can be turned off by imposing on the setup a magnetic field created by sending a control current through a nearby control wire—an ordinary current through an ordinary conductor. There you have it. When the control current is on, the switch is off, and when the control current is off, the switch is back on. It is an electromagnetically controlled superconducting tunneling current switch, another fine example of quantum physics technology at work!

The beauty of this triple super switch lies in its incredibly fast raw speed; some of the state-of-the-art experimental models have

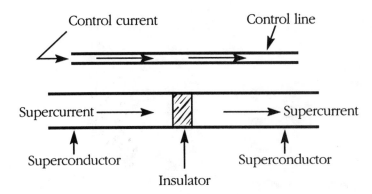

7-7 *Schematics of the Josephson junction superconducting switch.*

been clocked at a peak switching speed of about 10 trillion times per second. That is a speed of 10 terahertz, a thousand times faster than 10 gigahertz, and a million times faster than 10 megahertz. Some of the fastest supercomputers today run at a raw speed of a few gigahertz, and we spend a few thousand dollars when we wish to upgrade our personal computers from a low speed of 8 megahertz to a 35 megahertz. The power and the promise of microchips based on Josephson junction switches need no elaboration. Whoever first comes up with practical and widely commercial computers using these superconducting chips will soon claim the title to an entirely new generation of superfast computers, the superconducting supercomputers—another "SSC," not to be confused with the superconducting supercollider.

8

✳

The tunneling eighties

DESPITE ITS EARLY SUCCESS in nuclear radioactivity and its promise in microelectronics, the physical phenomenon of tunneling has largely remained an obscure secret of quantum mechanics. Some significant progress has been made over the years in the development of the Josephson junction device, but for reasons of technical impracticalities, the superconducting switches have yet to reach the status of full-fledged products. Until recently, quantum tunneling has been one of the least known facets of quantum physics.

The invention of the scanning tunneling microscope in the early 1980s began to change all that. With a magnifying factor thousands of times more powerful than the best electron microscope, the new device enabled us to see, for the first time, not only individual molecules but the contours of individual atoms. And as the first-ever picture of a gently twisting double helix of a DNA molecule has riveted our attention and fired our imagination, the obscure secret of quantum tunneling was suddenly thrust into the forefront of public view. As one breathtaking image after another of rows of atoms became widely disseminated, tunneling became almost a household word overnight.

Coming into the 1990s, the scanning tunneling microscope, or STM, performed another spectacular feat. Scientists discovered that the device can pick up an individual atom, move it a small distance, and deposit it in another location, one atom at a time, sort of like an atomic crane. In May 1990 scientists at the IBM Almaden Research Center in California did just that: They first scattered xenon atoms over a supercooled surface of a nickel plate, and then, using the tip of an STM, which is about one atom wide at its apex, picked up one xenon atom at a time moving and rearranging them until they formed the three letters "IBM," as shown in FIG. 8-1. The logo is made of 35 xenon atoms and is about 16 nanometers across, that is, about 160 angstroms wide. The average size of an atom is about 4 angstroms across, so at about 40 atoms wide, the sign might be the smallest logo ever "written."

Beyond the technology of scanning tunneling microscopes is another area in which the ultimate application of quantum tunneling is to play a crucial role. This is in the field of microelectronics. A bunch of electrons tunneling through a barrier represents an electric current, however miniscule it may be, and combined with

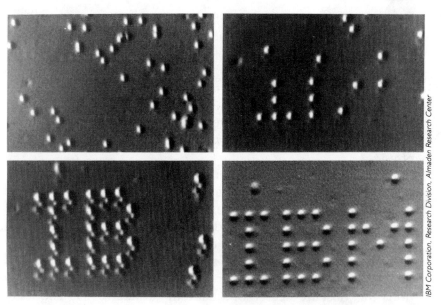

IBM Corporation, Research Division, Almaden Research Center

8-1 *The formation of an image using a scanning tunneling microscope. First xenon atoms were scattered in a nickel single-crystal surface (upper left). Then the xenon atoms were moved into position with the STM (upper right). The image at lower left appears blurry because the STM tip had three atoms at its apex instead of the desired one atom. The completed pattern (lower right) is shown magnified 2.9 million times.*

its sensitivity to the width of a barrier, this small amount of current provides the basis for building an electronic switch on an atomic scale. We have already met one such possibility in the form of the superconducting Josephson junction devices, but starting in the late 1980s, scientists began to focus on fabricating such tunneling switches out of ordinary silicon materials. Called quantum tunneling transistors, and still in their experimental stages, these devices pack so much power and processing speed into microchips that, when perfected, they could usher in a revolution in microelectronics, as profound and far-reaching as the invention of the original transistors did in 1947.

As the tiniest controllable current known, the quantum tunneling current is helping to redefine the very sense of the dimensionalities of technologies from those of a microworld to those of a nanoworld in which it is more appropriate to speak of nanotechnology, nanofabrication, nanochips, and nanoscopes. This chapter looks at these key areas of application of the tunneling effect, which had their beginnings in the decade just passed.

From microscopes to nanoscopes

Prior to the invention of the scanning tunneling microscopes, the notion of being able to see individual atoms, by any technique, was a distant dream. The smallest object that a then state-of-the-art electron microscope could distinguish was no smaller than about 5 nanometers, or 50 angstroms, long, and while such a power of resolution represented a thousandfold improvement over that of light-based microscopes, it was still much too coarse to tell one atom from another.

Why a scanning tunneling microscope can discern the kinds of detail that no other microscopes can is its radically different principle of operation. It does not "see" a sample by illuminating it, either with a beam of light as in the case of the ordinary microscopes, or with a beam of electrons as in the case of the electron microscopes. Instead, a scanning tunneling microscope maps the surface of a sample by letting the supersharp tip of its probe needle move vertically up and down, ever so delicately, following the contour of individual atoms. A scanning tunneling microscope does not so much "see" an atom as it "feels" it.

As the tip of a probe needle hovers over the surface of a sample, a controlling device keeps the gap at about 1 nanometer, just enough separation to maintain a desired level of tunneling current, by moving the tip up or down as it moves horizontally across the atomic landscape of the sample. These miniscule vertical movements are monitored by sensors and converted into corresponding electric signals, which form a computer-reconstructed image. This might come as a surprise to you, but this mode of operation—conversion of mechanical movements into electric signals—is as old and familiar as the way a stylus of an ordinary record player works. In both instances, the operation begins with converting slight movements of a needle into an analogous set of electric signals.

The way a record player cartridge is designed to do its job is schematically shown in FIG. 8-2. It is fairly simple and straightforward. The cartridge consists of three parts: a sharp diamond-tipped stylus, which is kept in gentle contact with the groove; a narrow strip of a metal spring, one end of which is attached to the stylus and the other end to the cartridge's third component; a piece of quartz, perhaps the most commonly used of a group of materials called piezoelectric crystals. These crystals possess molecular structures that produce proportionate electrical signals whenever a pressure is applied to them. Piezoelectric crystals not

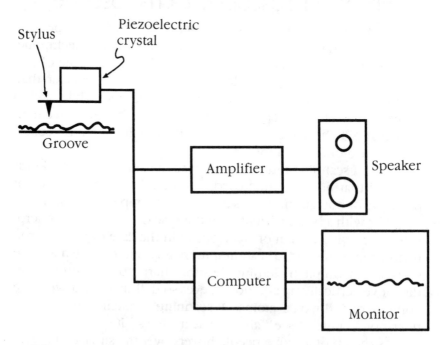

8-2 *A phonograph stylus tracking the contours of a groove.*

only sense variations in pressure, but also convert slight vibrations into electrical signals—or, conversely, applied signals into slight vibrations of crystals.

As a phonograph record turns, the stylus vibrates up and down following the contour of the microridges and microvalleys etched on a groove. The pattern of these vibrations is transmitted by the metal spring in a matching pattern of variations in pressures to the piezoelectric crystal, which then converts it into an analogous pattern of electrical signals. The rest is a matter of simple electronics: The feeble signals are pre-amplified, filtered, and screened for high fidelity and fed into an amplifier, which turns them into electric currents powerful enough to drive a set of speakers. Blasting away at their peak power output, the speakers perform their duty of delivering the maximum vibrations through our innermost bones!

If we wished to see a magnified picture of a groove and examine in detail every little bump and valley, it would be a relatively simple matter. We would channel the electrical signals coming out of a piezoelectric crystal not into a sound system, but instead into a computer graphics system, as shown at the bottom of FIG. 8-2. With the help of a relatively simple software program, the computer could plot on a display screen the three-dimensional image of detailed features of a groove. It would be a sort of computer-assisted electronic microscope. As the tip of a stylus tracks or "scans" a groove by touching the surface, its magnified images are reconstructed by a computer for us to see and record on videotapes. Such a system, a record player hooked up to a computer could be dubbed a "scanning touching microscope."

That is the idea behind the scanning tunneling microscope. Here, too, a piezoelectric crystal plays the crucial role of converting mechanical vibrations into electric signals, this time at a much higher level of precision. From these signals a computer system reconstructs three-dimensional images of bumps and valleys that correspond to individual atoms and molecules. What makes the scanning tunneling microscope unique—and this is what enables it to attain such powerful resolution—is that its business end, its "stylus," operates on the basis of the phenomenon of quantum tunneling.

The supersharp tip of a tungsten-needle probe is positioned so that the separation gap between the tip and the sample is narrow enough, but not actually in contact, for the electron waves to

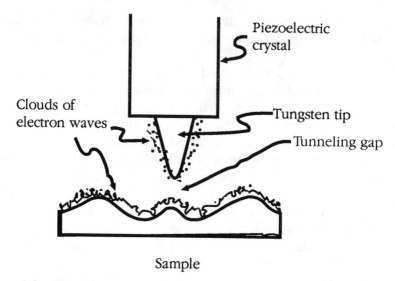

Piezoelectric crystal

Clouds of electron waves

Tungsten tip

Tunneling gap

Sample

8-3   *The "stylus" of a scanning tunneling microscope.*

tunnel across, as shown in FIG. 8-3. Owing to the wave nature of electrons at this scale of things, the tip of the probe is not exactly a point, but rather a collection of electron waves. Likewise, the outermost surface of a sample is not defined by a sharply drawn line, but rather by somewhat diffuse electron clouds, as indicated in FIG. 8-3 by a layer of dots representing the probabilistic distribution of electrons. When the gap between the electron clouds, those on the tip as well as on the surface of the sample, becomes narrow enough for some electron waves to bridge, tunneling begins. As the tip scans the surface of a sample horizontally, traversing one row of atoms after another, an extremely sensitive sensor monitors the resulting tunneling current and moves the tip up or down at each bump or valley of the atomic landscape in order to maintain a constant gap width and hence a constant amount of the tunneling current. Another sensor registers these miniscule vertical movements of the tip and piezoelectrically converts them into an analogous pattern of electric signals—just as in the case of an ordinary phonograph cartridge—which are then fed into a computer to be processed into images.

The tip of the tungsten needle is ground so sharp by an electrochemical etching process that in most instances the apex terminates on a few atoms. In some extreme cases it reaches the ultimate sharpness: a tip point consisting of a single tungsten atom. The gap of vacuum between the tip and the outermost surface of a sample is maintained as close as 1 nanometer, that is, about 10 angstroms, or barely wide enough to fit two average-sized atoms in a row. A minute voltage difference, typically a few hundredths of a volt, is maintained across the gap to enhance electron tunneling. In such a setup, the strength of the tunneling current is a few nanoamperes, or a few billionths of an ampere of electric current. This current permits tunneling at a rate of about 10 electrons per nanosecond.

As I explained in chapter 6, the probability of tunneling depends on the width of the gap. In a typical setup of a scanning tunneling microscope, the rate of tunneling is reduced by a factor of 10 for each angstrom of distance added to the gap. The rate of 10 electrons tunneling through per nanosecond across a gap of 10 angstroms would be drastically reduced to one electron per nanosecond tunneling through the gap if the width were increased to 11 angstroms. Sensing such a pronounced rise or drop in the tunneling current, a sensor would readjust the gap width by moving the tip up or down to restore the original gap width. This is the working secret of the scanning tunneling microscope.

The STM works best with solid metal samples. In solids, unlike liquid or organic molecules, atoms stay locked in their places in the lattice structure of solids. Also, a small voltage is maintained between the tip and the sample. As an STM scans across the surface of a microchip, the resulting image clearly shows the individual silicon atoms, grouped by their lattice patterns. Pictures of the honeycomb pattern of carbon atoms in a sample of graphite or the alternating rows of gallium and arsenic atoms in a gallium-arsenide semiconductor, where the interatomic spacings of about $1^1/_2$ angstroms are clearly visible, are just as breathtaking.

Images of the organic molecules of life obtained by an STM are not as sharp as those of solid metals because these molecules neither conduct electricity nor stay fixed in their places. They usually have to be either dried or frozen onto a sheet of metal before successful scanning can be done. Despite these apparent drawbacks—and these are being improved on constantly, this technology has produced some remarkable pictures: the first-ever image of the double helix of a DNA molecule, amino acid molecules that make up a protein as well as a DNA forming bonds to a protein. The technology has also made it possible to study the structure of the various molecules of life in great detail—structures previously only inferred by direct means—by bouncing X-rays off the molecules and studying the resulting patterns of their reflections.

STM technology is also playing an increasingly important role in our attempt to unravel the mystery of the recently discovered high-threshold superconductors by enabling us to understand the lattice structure of these copper oxide compounds. And, as we have seen, by increasing the voltage across the tip and the sample, an STM can be used as an "atomic crane," to pick up individual atoms, move them around, and redeposit them, thus opening up the possibility of manipulation of materials at the atomic level. The use of STM technology in semiconductors, superconductors, molecular biology, and biotechnology has only just begun, but already it has contributed additional evidence for the wave nature of matter in the quantum world.

Microchips, nanochips, and yes, tunneling chips

One of the latest developments in the application of quantum mechanics to today's high technology employs the finesse of the quantum tunneling effect to further reduce the microminiaturization of integrated circuits, thus pushing the frontier of nanofabrication technology down to the realm of quantum world. Beginning with tests in the mid-1980s and still in the experimental stage of development, the so-called quantum-effect integrated circuits can pack in so much more power, speed, and multifunctional versatility that they represent not just another generation of products but a truly quantum jump in microelectronics—or should I say nanoelectronics. Many scientists believe that these tunneling-based devices, when perfected in the near future, will usher in a dazzling new era of technology no less revolutionary than the invention of transistors more than 40 years ago.

An integrated circuit, or IC, is the original technical name given to a microminiaturized electronic circuit etched onto a tiny sliver of silicon—a complete self-contained circuit designed to perform a specific function. Better known by its nickname the chip, an IC is basically a circuit consisting mainly of transistors together with contacts and connectors fabricated on the surface of a silicon wafer, all on a greatly miniaturized scale. Although the transistors were invented back in 1947, the integrated circuit did not come into being right away. The technology for mass-producing transistors directly from thin flat wafers of silicon became available in 1957. Soon after, the world's first integrated circuit was invented by Jack Kilby at Texas Instruments and independently by Robert Noyes, who was then at Fairchild Semiconductor and who later founded Intel Corp. One of the first mass-produced chips is shown in FIG. 8-4. It was made of four transistors connected by about a half-dozen aluminum connectors, all fabricated on a piece of silicon wafer about 1/16 inch across. Compared to the megachips of today, the piece looks like some distant fossil from the Ice Age, but these crude chips ushered in the Age of Information as we know it today.

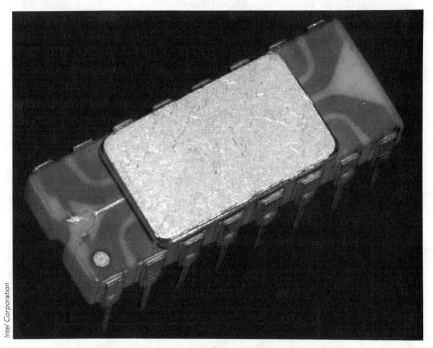

Intel Corporation

8-4 *One of the earliest integrated circuits.*

Throughout the 1960s the technology of downsizing transistors made steady and gradual progress, but not until the early 1970s did integrated circuits containing thousands of transistors—referred to as LSIs for large-scale integrated circuits—begin to be mass produced. In 1971, a full decade after the invention of the original chip, Ted Hoff at Intel designed a completely self-contained computing unit fabricated on a single tiny sliver of silicon. Named the Intel 4004 chip and containing 2,000 transistors, it was the world's first complete logical and arithmetic processing chip, the so-called computer-on-a-chip, which has since come to be known as the microprocessor. Within the next three years, the 4004 chip was superseded by the new Intel 8080 microprocessor, which contained 5,000 components and could add two numbers in less than 3 millionths of a second. By the end of the 1970s, with the invention of the Apple II personal computers—touting the then "awesome" memory of 4,000 characters— and the newer Intel 8086 microprocessor becoming an industry standard, the age of microcomputers—the PC revolution—was up and running at full steam.

The pace of progress quickened coming into the 1980s. The scale of microminiaturization kept shrinking, and the number of components crammed into a single chip increased exponentially. With that the power and speed of microprocessors shot straight up. Gone were the days of LSIs, replaced by the decade of VLSIs, or very-large-scale integrated circuits, chips that contained hundreds of thousands and eventually millions of transistors. By the time IBM brought out its original PC in 1981, a superchip was unveiled by Hewlett-Packard that took a team of engineers two years to design and was built out of a half-million components. Around this time, too, two companies established themselves as dominant in the microprocessor industry: Intel, with its dynasty of 80 x 86 microprocessors—the 80286 in 1983, the 80386 in 1986, and the 80386SX and 80486 microprocessors in 1989—and Motorola, with its line of 68000 chips—the 68020 in 1984, the 68030 in 1987, and the 68040 microprocessor in 1990. Intel's i486 microprocessor packs 1.2 million transistors into an area approximately 1/2-inch square.

Some of the most advanced chips available in the market today contain as many as 10 million components—transistors, capacitors, and connectors—in microscopic dimensions, all packed in a tiny sliver of silicon no larger than a thumbnail. A state-of-the-art memory chip, the so-called 4-megabit dynamic random access memory (DRAM) chip capable of storing 4 million on-and-off signals, is one such VLSI chip, but already engineers are experimenting with prototypes of much higher-density chips containing up to a quarter-billion components. The idea of manufacturing electronic components that small might appear incredible, but these objects are within easy reach of our direct comprehension, and, when compared to the typical dimensions of the quantum world, they are still quite large.

The unit of length most often used to describe a VLSI chip is the micron, a millionth of a meter. The smallest unit of length we are most familiar with in our daily routine, meanwhile is the millimeter. Any ordinary 12-inch ruler has one edge calibrated in inches, usually down to 1/16 inch, and a metric calibration on the other, 1 millimeter being the smallest marking. Some comparative scales in the world of microchips are shown in TABLE 8-1. With a little help, say by holding a magnifying glass over a ruler, it is not difficult for the naked eye to make out an object down to the scale of one-tenth of a millimeter, or 100 microns, which is the average width of a human hair. In most commercially available general-purpose chips of today, the transistors are one to two orders of magnitude smaller than the width of a human hair, ranging from several microns down to one micron. At the larger end, a transistor is roughly the size of a human red blood cell. At one micron, one-hundredth the size of a human hair, it has about the same dimension as interstellar dusts particles. The lateral spacings between components in a 4-megabit memory chip are about one micron. One micron is about the lower limit for present-day chip fabrication technology, which is often referred to as micron technology. However, experimental chips have been produced for the U.S. Department of Defense in which the transistors are as small as half a micron.

Table 8-1   *Some benchmark scales in the world of microchips.*

1000 microns = 1 mm	The smallest unit of length marked on a standard 12-inch ruler.
100 microns = 0.1 mm	The width of a human hair.
10 microns = 0.01 mm	The size of a human red blood cell.
1 micron = 0.001 mm	The size of the smallest transistor used in commercially available chips today, one-hundredth of a human hair's width.
0.1 micron = 0.0001 mm	One-thousandth of the width of a human hair, the most likely limit for the smallest transistor that can be manufactured by the conventional technology.

To get a better feel for the micron-sized objects in terms of things more familiar and realistic, consider the following two examples: Suppose you draw on a piece of paper a square precisely 1 centimeter by 1 centimeter. The small square is almost exactly one-quarter the size of a standard postage stamp. Since we can readily make out, with the help of a magnifying glass, dimensions down to one-tenth of a millimeter, it is not difficult to visualize having drawn 100 lines down and 100 lines across on this little square. That is 10,000 intersections in all. Now let us go a step further, that is, one order of magnitude smaller, and imagine that we have drawn 1000 lines down and 1000 lines across on this patch of an area about a quarter of a stamp in its size. The gap between two adjacent lines is now one-thousandth of a centimeter, or 10 microns. To design an integrated circuit by placing one tiny transistor at every intersection, we would have a chip that contains 1 million transistors. If these transistors were each about 6 microns across, the lateral gap between transistors on this "stamp" would be about 4 microns wide. A VLSI chip made of a few million parts is not really as far-fetched as it might have seemed.

The second example has to do with a microfiche, the flat sheet of microfilm that stores miniature photocopies of printed materials for a library, archive, or even a warehouse or auto parts shop. Most of the time, we just read them—or watch someone else struggling with them—without paying much attention to the reduced dimension of letters on the sheet. A standard sheet of microfiche is about 4 by 6 inches in size and typically contains the miniature photocopies of about 100 letter-size pages. For the ease of a quick estimate, let us say an original document page is 8 by 10 inches and each microfiche contains copies of 96 such pages, eight pages down the side 4 inches long and 12 pages across the side 6 inches long, that is, 80 inches of the original pages reduced to 4 inches on the microfiche. It is a 20-to-1 miniaturization, not such a big deal. Now take out a ruler and measure the diameter of the period at the end of this sentence on this page—or, better yet, measure the gap between the dot and the stem of the letter "i." Let's say the period measures about two-tenths of a millimeter, that is, 200 microns. Downsized by a factor of 20-to-1, the period corresponds to about 10 microns on a microfiche, and that is about the size of a relatively large transistor on today's VLSI chips. No one gets terribly excited or awed by a simple run-of-the-mill microfiche because it is not such a big deal. Neither is a VLSI chip containing a mere few million transistors.

Fabrication of commercial components at a scale smaller than 1 micron, appropriately called submicron fabrication technology, is one of the most hotly pursued goals of the semiconductor industry today. In the summer of 1990 a prototype of a 64-megabit memory chip was successfully tested, and it contained parts as small as 0.35 microns. An experimental transistor about 0.1 micron in size, or about one-thousandth the thickness of a human hair, is known to have been successfully made in laboratories. The wavelengths of the visible light spectrum, by comparison, range from 0.7 micron for the red to 0.4 micron for violet. In our unending quest for more powerful, versatile, and faster computers, we would like to push the threshold below even the 0.1 micron range. Unfortunately, at this level we begin to bump into serious technological limitations, at least with the conventional technology of today. Even if we could overcome these limitations, we still face the formidable barrier imposed by the laws of physics: Somewhere between the dimensions of 0.1 to 0.01 micron we come to the threshold that separates the ordinary world of Newtonian physics from the realm of quanta where the wave-particle duality of matter begins to assert itself.

Transistors inside an integrated circuit serve as simple on-and-off switches—gates that either block or pass an electric current—in a system of millions of switches. The on-and-off sequences of electric currents correspond to the physical form of binary language, strings of 0s and 1s, with which the machines do their thing. As the size of these transistors shrinks, down to the 0.1 micron range, the gaps of insulating material between them also become narrower, so narrow that eventually electrons from one transistor begin to tunnel across the barrier to adjacent transistors. What we have is short circuiting by a quantum wave, a matter wave jumping tracks. Clearly such current leakage cannot be tolerated, and this is the physical limit imposed by the laws of quantum physics on the technology of microminiaturization. By some estimates, this limit will be reached by conventional chipmaking technology before the end of this decade.

Undaunted by what at first appears to be an unsurmountable physical limit, however, scientists have launched an ambitious new research program aimed at enabling them to overcome even this quantum physical limit. The idea behind it is both simple and bold: Why not turn the problem of ''leakage'' caused by electron tunneling into the mechanism for building a switching device? The ''on and off'' of it would then correspond to whether electrons tunnel or not. As in the case of the scanning tunneling microscope, a minute amount of tunneling current can be used to provide the basis for a totally new breed of transistors—devices appropriately dubbed "quantum tunneling transistors." If successful, this microminiaturization process can be pushed right into the dimensions of the quantum world, and the number of transistors that can be packed into a single chip would increase a millionfold and, accordingly, so would the power and speed of the chips of the future.

Still in its embryonic stage of development, the basic workings of a quantum tunneling transistor—also called by such other names as a quantum well or a resonant tunneling transistor—can be depicted as shown in FIG. 8-5. A thin layer of a semi-insulating material forms an electrical barrier for the conducting electrons between semiconductors, such as silicon or compounds of gallium arsenide or indium phosphide. The gap is typically about 0.03 to 0.02 micron across and can be made as narrow as 0.01 micron, narrow enough for the electron waves to negotiate the gap and tunnel across it from one side to the other under a favorable condition. As shown in chapter 6, the energy profile that the barrier presents to oncoming electrons can be represented as shown in the figure: Whether an electron will tunnel through the barrier or not depends on the width of the energy profile at the level at which the electron strikes it.

A tiny amount of a control signal current to the semi-insulating barrier will effectively raise or lower the energy profile of the

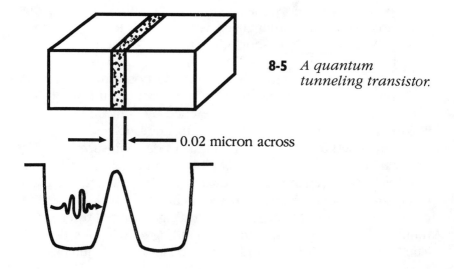

0.02 micron across

8-5 *A quantum tunneling transistor.*

barrier depending on whether the applied signal is positive or negative. As depicted in FIG. 8-6, the energy profile of a barrier that is wide enough to block tunneling can be lowered just enough in this manner to let an electron tunnel across it. That is what a switch should do—and what ordinary transistors do inside chips every nanosecond of the day. The new devices could perform this switching function much faster than today's transistors consuming a lot less energy and generating a lot less heat, two prerequisites for the super-hyper supercomputers of tomorrow.

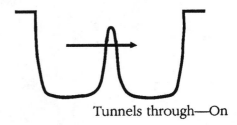

Tunnels through—On

8-6 *The on and off of a tunneling transistor.*

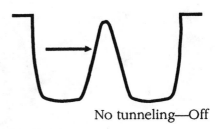

No tunneling—Off

This ambitious approach faces a long list of technological challenges that must be overcome before it achieves commercial viability. The quantum dimensions involved in making these devices render them extremely sensitive and susceptible to external influences. While tunneling is allowed under a tightly controlled environment inside a transistor, no tunneling can take place between transistors. Extensive shielding of chips will be necessary and that includes shielding from any form of electromagnetic interference that could raise havoc inside such a chip—anything from sparks from a neighbor's garage to the shower of powerful particles in cosmic rays such as photons, neutrinos, and protons. Some of these technological problems are formidable. But when these quantum tunneling transistors are fully developed, as surely they eventually will be, and become commercially available, microprocessors and memory chips built from them will be so much more powerful, faster, and energy-efficient that future computers—super quantum tunneling computers, that is—will make today's fastest multimillion-dollar supercomputer look like a child's toy. At least that is what *can* happen in the near future, with the help of the wave-particle duality of quantum mechanics.

✳

Epilogue
Quantum field theory and the wave-particle duality

Some 90 years since the original inception of the idea of quanta, we have come to witness the most direct technological application of one of the arch examples of wave-particle duality: the tunneling by matter waves across impenetrable barriers. Since its completion in 1930, quantum theory has been spectacularly successful in explaining new facts about the physical world—not only of atoms and atomic nuclei, but also of elementary particles and the way they interact with each other. It is refreshing to observe that, after all these years, a development of a cutting-edge technology has taken us right back to the basics of quantum mechanics: an adaptation of quantum tunneling, as a mechanism for such a basic task as switching on and off an electric current.

The theory of quantum physics itself has gone through several stages of evolution during the past 60 years: First was the formulation of the relativistic counterpart of quantum mechanics, or, if you like, the quantum counterpart of Einsteinian mechanics, the quantum mechanics appropriate to the world of the small and fast. This merging of the two branches of modern physics resulted in a theory to describe the world of tiny elementary particles moving about with great speeds and is called, appropriately enough, relativistic quantum mechanics. This was soon followed by the formulation of the quantum counterpart of classical electricity and magnetism, the quantum theory of the interaction between matter and radiation in the world of the small, the so-called quantum electrodynamics. By the end of the 1940s, both the relativistic quantum mechanics for matter particles such as electrons and the quantum theory of electricity and magnetism had been incorporated into a general theoretical framework. This grand synthesis of waves, particles, and forces in what is called the relativistic quantum theory of interacting fields, or quantum field theory for short, remains to this day the most advanced theoretical formalism for fields and particles that physicists have been able to construct. Let's briefly trace this evolution.

At the beginning of this book, in chapter 1, I discussed the limits beyond which classical Newtonian mechanics was no longer valid. One such limit was in the realm of speeds; for the world of the very fast, the old mechanics had to be replaced by the new relativistic mechanics of Einstein. The other limit, which necessitated the formulation of quantum mechanics, was in the dimension of size. Specifically, I used simple block diagrams in FIGS. 1-1 and 1-2 to represent these relationships. As you glanced at the figures, you likely had a feeling of something incomplete, a sense of a missing block that should have logically been there to complete the picture, but of which nothing was said. The "missing" block is the void at the intersection of the quantum and relativistic mechanics, which, when filled, completes the picture as shown in FIG. E-1.

	Slow ⟵⟶ Fast	
Large ↕ Small	Classical mechanics	Relativistic mechanics
	Quantum mechanics	Relativistic quantum mechanics

E-1 *The missing block.*

Relativistic quantum mechanics, as it is most appropriately called, was developed by Paul Dirac (1902–1989), who constructed an equation for a matter wave—specifically for an electron—similar to the wave equation by Schrödinger but consistent with the relative nature of space and time. The equation, known as the Dirac equation, was completed in 1928, just two years after Schrödinger came up with his in 1926. The two equations, those of Schrödinger and Dirac, form the mathematical centerpieces of quantum mechanics.

In addition to being the relativistic extension of quantum mechanics, this new relativistic theory contained one important prediction: the existence of antimatter in the universe. Working from his equation, Dirac came to realize that relativistic quantum mechanics could not be complete without the existence of a particle that was a mirror image of an electron, a particle identical to an electron in every respect save one: Its electric charge had to be equal in magnitude but opposite in sign from that of an electron. This then, in 1928, was the beginning of our knowledge of the existence of antimatter. The newly predicted particle, an anti-electron, was soon named a positron after its positive electric charges. When a positron was first discovered in 1932, it confirmed not only the existence of antimatter but also relativistic quantum mechanics itself. For their towering contributions, Schrödinger and Dirac shared the Nobel physics prize in 1933.

Classical mechanics is one of the two great physical theories that constitute classical physics, the other being the classical theory of electricity and magnetism, or classical electrodynamics. If mechanics needed such fundamental revisions when extended into the relativistic and quantum domains, wouldn't it also be the case for electrodynamics, as some of you might have wondered, requiring a drastic revision of its own? The short answer is no and yes. No, the relativistic modification of it is not needed at all, but, yes, a new extension of it to deal with the behavior of electrodynamics in the world of photons is definitely needed, namely the quantum counterpart of classical electrodynamics—or the electrodynamic counterpart of quantum mechanics, quantum electrodynamics. The block diagram for electrodynamics is thus simpler, as shown in FIG. E-2.

The fact that electrodynamics, the theory of electricity and magnetism developed over a period of about 100 years roughly from 1780s to late 1880s, actually required no fundamental revi-

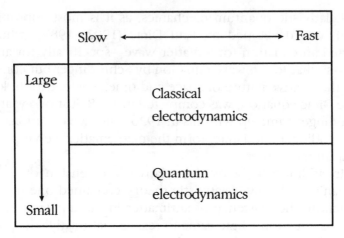

E-2 *The large and small of electrodynamics.*

sion when extended into the relativistic domain might at first come as a surprise. It surprised many experts at the time. A little reflection should convince you, however, that for a theory that contained the equation for the propagation of light itself no fundamental revision should really be necessary to come to terms with the theory of relativity, one of the founding principles of which was the absolute constancy of the speed of light in vacuum. As I mentioned in chapter 3, when in 1864 Maxwell achieved the grand synthesis of the laws of electricity and magnetism into a set of four unifying equations, called Maxwell's equations, the resulting theory predicted the existence of the electromagnetic wave, of which light was a part, and the speed of its propagation, which in vacuum has an absolute unchanging value. All the uniquely singular properties of light were already fully contained in the theory. In other words, the theory of classical electrodynamics completed by Maxwell in 1864 was already fully relativistic. We just didn't know it until Einstein came along in 1905.

Extending the laws of electricity and magnetism into the domain of quanta—the microcosm of photons, electrons, and positrons—was another matter. Quantum mechanics, relativistic or otherwise, is, after all, the wave theory of matter, a wave theory of particles. Although it was the discovery of photons, particles of light, by Planck and Einstein that started the whole ball rolling, a proper theory for photons themselves first had to be formulated. What was needed was a particle theory of a wave; a fully quantum theory of photons for the electromagnetic radiation had to be formulated. Such was quantum electrodynamics, the quantum counterpart of classical electrodynamics. As it turned out, the task of formulating quantum electrodynamics, or QED, was much more complicated mathematically than even quantum mechanics itself. Building on the success of relativistic quantum mechanics as put forward by Dirac, many physicists toiled at this complicated task. The theory of QED we know today was completed by the early 1950s, largely at the hands of three great theorists: Richard Feynman, Julian Schwinger, and Sinitiro Tomonaga. The trio shared the 1965 Nobel prize in physics for their contributions.

Quantum electrodynamics remains the most successful theory of its kind to this date: Not only does it provide a consistent theoretical framework for the existence of photons, exactly the type of a particle theory of a wave, but also, going much further than that, it is a complete theory for the workings of the electromagnetic force at the quantum level. QED has been spectacularly successful in explaining just about every aspect of electromagnetic interaction in the world of photons, electrons, and anti-electrons: the way they interact with each other, electrons with electrons, electrons with positrons, and positrons with positrons, as well as how matter and antimatter (in this case, electrons and positrons) can be created by or disappear into electromagnetic radiation. It is a powerful theory. Its results agree with actual observations and data better than one part in 10 million. To this day, we have not encountered a single experimental instance in which a serious disagreement between the theory and experiment has been recorded.

For all its power and success, however, the details of QED form a subject that is extremely esoteric, known only to a small fraction of professional physicists. For one thing, QED claims a fame of another sort: Its renowned mathematical complexity and the method of calculations are both lengthy and tedious. It takes a considerable amount of mathematical dexterity to hack a path through the subject and even more to carry out detailed calculations—often using strange-looking charts known as Feynman diagrams. A majority of physics students pursuing their doctoral degrees never have to take a course in quantum electrodynamics.

Hidden under the mathematics is a fundamental conceptual contribution of quantum electrodynamics that provides a basis for interpretation—if not an outright complete explanation—of the founding rock of all quantum physics: the wave-particle duality. In electrodynamics the wave in question is the electromagnetic wave, a narrow portion of which is visible light, formed by fluctuating electric and magnetic fields. In the theory of quantum electrodynamics, the field energy, or the energy carried by the electromagnetic field, cannot have any old value in a continuous spectrum of values. It can only have a discontinuous and discrete set of values, that is, a countable multiple of a basic and indivisible unit of energy. This indivisible unit of energy, in the case of light, is the photon, the quantum or particle of light. In other words, a field is a wave whose energy can have values only in an integer multiple of the energy of its quantum, the particle corresponding to the field. That a physical object—be it an electron, a positron, a proton, or even a single atom—can be thought of as both a particle and wave, in the weird realm of quanta, has found a mathematical interpretation as the duality of a field and its quanta, the field as the wave and its quanta as the particles of that field.

The fact that the energy of the electromagnetic field comes in discrete packets of photons was not, of course, something advanced by quantum electrodynamics; it was the original empirical discovery of Planck and Einstein. What the theory of QED achieved was to provide a self-consistent theoretical framework in which the wave-particle duality could be mathematically expressed. A field might contain a few thousand, a few hundred, or only several of its quanta, and in the extreme consists of a single quantum. A field interacts with others by transferring its energy only in discrete amounts of its quanta, each of which acts as a particle imparting its energy as an indivisible whole. In this picture of duality, the interpretation of duality in terms of a field and its quanta, the creation or annihilation of particles corresponds mathematically to an act of raising or lowering the energy of the field.

To be sure, the mathematical relationship between a field (wave) and its quanta (particles) afforded by quantum electrodynamics can no more explain the basic mystery of the duality—that it spreads like a wave, but impacts like a particle—than ordinary quantum mechanics. But such a relationship does provide a convenient mathematical handle to pursue the matter further. This field-theoretic viewpoint of the wave-particle duality became quickly established as one of the guiding principles for all fields and matter. Just as a photon is the quantum of the electromagnetic

Fields (wave)	Quanta (particles)
Electromagnetic field	Photons
Electron field	Electrons
Positron field	Positrons
Proton field	Protons
Neutron field	Neutrons
•	•
•	•
•	•

E-3 *The field-quanta duality of quantum field theory.*

field, an electron becomes the quantum of an electron field, which is the de Broglie's matter wave for an electron. Likewise, a proton is the quantum of a proton field, a neutron the quantum of a neutron field, and so on. Every known field is associated with a particle, its quantum; conversely, every known particle is associated with a field for which it is the quantum. Trimmed to its barest bones, this is the underlying reasoning for the quantum field theory: the wave-particle duality expressed as a mathematical relationship between a field and its quanta.

The wave-particle duality, as the very defining characteristic of the quantum world, has come a long way over the years, from an inexplicable experimental discovery in 1900 to a lofty mathematical description in quantum field theory. In the intervening years, quantum physics ushered in the modern science and technology of the twentieth century—everything from molecular bonds to quarks, from semiconductors to supercomputers, and from the age of machines to the age of information. Despite all this, however, quantum mechanics, which began life with puzzles and paradoxes, remains a riddle to this day. To quote Feynman once again: "I think I can safely say that nobody understands quantum mechanics." The riddle remains.

Glossary

alpha decay A nuclear emission process in which an unstable heavy nucleus emits a high-speed alpha particle and transforms itself into a different and lighter nucleus. This process is one of the three processes, together with the beta and gamma decays, that make up nuclear radioactivity. Despite the name "decay," nothing is rotting away here.

alpha particle A helium nucleus that comprises two protons and two neutrons; it is a tightly bound system. Played a crucial role in experiments in 1911 that revealed the planetary structure of atoms.

angstrom A unit of length used characteristically in the world of atoms, being equal to one-tenth of one-billionth of a meter (10^{-10} meters), or about four-billionth of an inch. Radii of atoms range between 1 and 3 angstroms. A nanometer, a computer-age unit of length, is 10 angstroms.

antiparticle Each particle has a corresponding antiparticle that has the same mass but equal and opposite electric charges. A proton has a negatively charged antiproton and an electron a positively charged antielectron, called a *positron*. The existence of antimatter was first predicted by Paul Dirac in 1928 on the basis of the relativistic version of quantum mechanics that he formulated. At the more fundamental level, every member of the basic building blocks of matter, now believed to be three distinct sets of families of quarks and leptons, is associated with its own antiparticle that has not only equal and opposite electric charges but, more generally, equal and opposite charges with respect to the strong nuclear force (the color charges) and the weak nuclear force (the weak charges). An antineutron is made of antiquarks and is distinct from a neutron. If an antiproton and a positron were to form an antihydrogen atom, that atom would be distinct from a hydrogen atom even though both are electrically neutral. Antineutrinos are opposite to neutrinos with respect to weak charges.

beta decay   One of the three nuclear emission processes called nuclear radioactivity, in which an unstable nucleus goes through a transmutation, emitting an electron with an accompanying antineutrino. This is the work of the weak nuclear force. In some processes, a positron is emitted instead with an accompanying neutrino. Positrons emitted from such beta decay processes are used in a state-of-the-art medical imaging technique called *positron emission tomography* (PET).

charge   Traditionally, electric charges, the sources of the electromagnetic force. The word came to be used as a generic term for the sources of all the forces, such as the color charges for the strong nuclear force. In this generalized sense, the masses of matter are gravitational charges.

chip   A popular term for an integrated circuit, the microminiaturized electronic circuit, no larger than a baby's fingertip, containing millions of transistors and other electronic components fabricated out of semiconductors; also known by such names as megachips, nanochips, and superchips.

color charge   Basic charges of quarks that act as sources for the strong nuclear force. In much the same way that electric charges come in two types, positive and negative, color charges come in three types, whimsically named red, green, and blue. The color charge force among quarks gives rise to the nuclear force among protons and neutrons. The theory of the color force is called *quantum chromodynamics* (QCD) in the same manner that the theory of the electromagnetic force is called *quantum electrodynamics* (QED).

electromagnetic force   One of the four basic forces in nature, which include gravity and the strong and weak nuclear forces, it is the force that acts between electric charges and currents. It is responsible for the structure of atoms and molecules, but inside a nucleus its repulsive force between protons works against the strong nuclear force, thus resulting eventually in nuclear instability. In the so-called Standard Model, the current theory of particles and forces, a mathematical framework has been advanced in which the electromagnetic force and the weak nuclear force are considered different manifestations of the same force— the *electroweak force*. Many details of this "unification"remain to be verified.

electromagnetic radiation   A wave of electric and magnetic fields that propagates through space at the speed of light, carrying its own massless form of radiation energy. Its spectrum covers radio waves, TV waves, microwaves, infrared waves, visible light, ultraviolet rays, X-rays, and gamma rays. Unless the existence of a gravitational wave can be confirmed, electromagnetic radiation is the only wave of its kind in the universe.

energy   A measure of the capability, endowed by a force applied to it, of matter to be able to do things such as move, accelerate, exert forces on others, or heat things. A force sets matter into motion, giving it a kinetic energy. A force might relocate a heavy object to a higher position, giving it a potential energy. At sufficiently high energies, masses can be converted into electromagnetic energy and vice versa.

fermi   A unit of length used exclusively in nuclear physics, being equal to 10^{-15} meters, that is, one-millionth of one-billionth of a meter. It is named after Enrico Fermi, who in 1942 built the first successful self-sustaining nuclear fission reactor on the campus of the University of Chicago. Nuclear radii range between 1 and 7 fermis. That 1 fermi is equal to 100,000 anstroms serves to illustrate the relative proportions of sizes between atoms and their nuclei.

field   In principle, a field is any quantity that is defined over an extended region of space, but the term is most often used in connection with influence of long-range forces spread over an extended space and time, such as the electric field, magnetic field, and gravitational field. In quantum theory, however, a field represents the very "waveness" of a particle, such as an electron field or proton field.

frequency   The number of cycles per second, 1 hertz being the unit for one cycle per second. For a moving wave, the number of crests passing through a given point per second; for a generator of electrical pulses, the number of on-and-off cycles per second. A computer with a raw speed of 25 megahertz has pulses turning on and off at the rate of 25 million times per second.

gamma decay   The third state of nuclear radioactivity, in which a nucleus makes a transition from a higher energy level to a lower one by emitting a high-energy photon. Photons emitted in nuclear gamma decay are more energetic than photons of visible light. The fact that this form of radioactivity is an emission of electromagnetic radiation sometimes lends itself to an easy confusion between radiation and radioactivity.

gigahertz   One gigahertz corresponds to the frequency of one billion cycles per second. It is a benchmark speed for some of the fastest supercomputers of today. At the rate of 1 gigahertz, a single on-and-off pulse corresponds to a time slice of 1 nanosecond.

high-temperature superconductor   Abbreviated HTS, it refers to a group of copper oxide materials recently discovered to become superconducting at temperatures relatively higher than previously thought possible. Until the discovery of HTS in 1986, the highest temperature recorded at which superconductivity began was just below the boiling point of liquid neon, −409 degrees Fahrenheit. Some types of HTS display superconductivity beginning at a temperature as high as −243 degrees Fahrenheit. All the excitement about the discovery of HTS swirls around the potential application of superconductivity at the even higher temperatures of relatively abundant and inexpensive liquid nitrogen.

integrated circuit   Also called a *chip*. An entire electronic circuit, even a whole computer with microminiaturized transistors and capacitors, can be fabricated on a single piece of a semicondutor material, such as silicon or gallium arsenide. Typically a rectangle less than 1/2 by 1/2 inch, it is a marvel of today's high technology.

interference   A term used to characterize a superimposition of many wave trains of different amplitudes, frequencies, and wavelengths, resulting in a wave of a completely different shape. The most dramatic interferences occur when two identical waves are superimposed either in phase (crests on crests) or completely out of phase (crests on troughs), resulting in a doubling reinforcement or a complete cancellation. Called constructive and destructive interferences, respectively, they provide the simplest, most direct, and unassailable tests of confirmation of a wave.

Josephson junction   An ultra-fast switching device made from superconducting materials. A thin insulating barrier is sandwiched between two superconducting strips in such a way that by allowing or blocking a quantum tunneling process to occur, it can serve as a fast and sensitive switching device. When fully developed for commercial use, it promises to increase the speed of computers more than a thousandfold.

large-scale integration Abbreviated LSI. The technique of fabricating an integrated circuit with thousands of electronic componants. Today it is more common to talk about a VLSI, a *very-large-scale integration* chip that contains a few million transistors, such as the 4-megabyte memory chip.

laser An acronym for *light amplification by stimulated emission of radiation*. A beam of light at a single frequency, hence one color, that travels in one direction and, most importantly, contains only coherent waves, that is, waves that move and fluctuate in perfect unison—crest-to-crest and trough-to-trough.

lepton A group of relatively light particles that, together with a group of quarks, constitute what we now believe to be the basic building blocks of all matter. A basic difference that separates leptons from quarks is that leptons do not participate in the strong nuclear force reactions. There are currently thought to be three distinct, but similar, families of leptons, the most familiar family being that of electron and its ''electron-type'' neutrino. The other two are muons and their associated neutrinos, and tau particles (the so-called heavy leptons) and their associated neutrinos. Just precisely how one family differs from the other, other than their different masses, is an unanswered question.

matter wave In the microcosm of atoms, within the reduced dimensions of the quantum world, physical objects, whether radiation or matter, behave in a mysterious way by acting sometimes as a particle and at others as a wave. The ''waveness'' of what had been considered solid particles, such as electrons and protons, is generally referred to as the matter wave. The idea was first postulated by de Broglie in 1924, and the matter wave is sometimes called the de Broglie wave.

memory chip A chip containing thousands and frequently millions of electronic cells, each storing a single bit of information by being either on or off. A *random access memory* (RAM) chip stores bits temporarily during the processing of a program, and loses all memory when turned off. A *read-only memory* (ROM), however, is a permanent storage for data and instructions.

microcomputer   A computer based on a *microprocessor*. Includes all personal computers and workstations. Ordinarily any desktop or stand-alone computer is a microcomputer.

microprocessor   A central processing unit built on a single chip. Sometimes called a computer-on-a-chip.

modern physics   As distinguished from classical physics, which consists mainly of Newtonian physics (mechanics), the classical theory of electricity and magnetism, and thermodynamics. Modern physics is the physics of the twentieth century, consisting of two revolutionary theories: relativity and quantum mechanics.

nanosecond   One-billionth of a second, a common unit of time used to measure the processing speed of faster computers, such as executing one instruction in so many nanoseconds. The speed of light in vacuum is 1 foot per nanosecond.

nanotechnology   A relatively recently coined term that covers the emerging technology, especially in electronics and materials science, of fabrication of superminiaturized components literally in a nanometer (billionth of a meter) scale.

nucleon   The generic name for the charged as well as the neutral constituents of atomic nuclei; that is, protons and neutrons. Setting aside the electromagnetic difference, protons and neutrons behave in the same way with respect to the strong nuclear force.

neutrino   The lightest member of a lepton family. Neutrinos are perhaps the strangest of all particles observed to date. They have no mass, no electric charge, and no color charge. They exist by the sheer virtue of the weak nuclear force alone and zip across the void at the speed of light, yet come in three different species: the electron-type, the muon-type and the tau-type. Just exactly how each is different from the others is still a mystery.

photon   A quantum of radiation energy, the particle of light. The first example of the particles of a wave, in this case the electromagnetic wave, the flip side of which is the matter wave of a particle. It defines the smallest indivisible unit of energy for a radiation of a given frequency. The energy of a photon is proportional to the frequency of the radiation of which it is the quantum. Although it is imperceptibly small by our standards, the spread of the energies of photons corresponding to different frequencies can be enormous. The energy of a single photon of radiation at a frequency of 1 terahertz (10^{12} cycles per second) is a million times greater than that of a single photon of frequency of 1 megahertz (10^6 cycles per second). In principle, mathematically speaking, the energy of a single photon of radiation of an infinite frequency would be infinite.

positron An antielectron that has exactly the same mass but equal and opposite electric charges as an electron. Positrons and electrons annihilate each other when they meet and turn into radiation. This is the basic mechanism behind the medical imaging technique called *positron emission tomography*, or PET.

probabilistic interpretation The mathematical interpretation of the wave-particle duality of the quantum world, which states that the amplitude squared, at a given position, of a matter wave corresponds to the probability with which the matter will be observed as a particle at that location. Combined with wave-particle duality, it is perhaps the strangest of all founding principles of quantum mechanics.

quantum An elemental unit of energy, originally of the electromagnetic radiation, that is, the photon. The concept has since been generalized to apply to all known particles, as in an electron being the quantum of an electron field, a proton being the quantum of a proton field, and so on.

quantum electrodynamics The theory of the electromagnetic force in action at the level of the quantum world, the world of atoms, electrons, and photons.

quantum mechanics The mathematical theory governing the physical behavior of the atomic and subatomic world. It is the only physical theory that we know that deals with the world of atoms, atomic nuclei, elementary particles, quarks and leptons, and fundamental forces of nature.

quantum tunneling transistor Also known by such other names as *quantum effect transistor, quantum effect device,* or *quantum resonant-tunneling transistor.* This is one of the latest cutting-edge devices in microelectronics, and is still in the experimental stage. When fully developed, these devices are expected to be so small and yet so fast and powerful that they would completely revolutionalize the world of computers.

quark   A fundamental particle that carries the color charges of the strong nuclear force and makes up protons and neutrons. Just as leptons come in three families of two each, quarks also come in three families of twosomes: "up" quark and "down" quark making up the doublet of the first family, "charm" and "strange" quarks making up the second family, "top" and "bottom" quarks, the third. Quarks are the source particles for the strong nuclear force. Despite an impressive accumulation of indirect evidence, so far no quark has ever been observed in isolation. Existence of the top quark, even in terms of indirect evidence, has not yet been demonstrated. One of the overriding purposes of the *superconducting supercollider* is to unravel the deep mysteries surrounding quarks.

radioactivity   A generic term for the three nuclear decay processes: the alpha, beta, and gamma emissions.

relativistic quantum mechanics   The extension of quantum mechanics into the realm of high speeds, that is, the relativistic speeds. A theory in which Einsteinian relativity and quantum mechanics have been merged to describe the world of highly energetic elementary particles. One of the important outcomes of this theory was the prediction, and its subsequent confirmation, of the existence of *antimatter*, the antiparticles.

scanning tunnel microscope   A revolutionary new microscope that operates on the quantum principle of electronic tunneling, a phenomenon that has no analog in our human-sized world. The STM is capable of resolving objects down to the scale of one-hundredth the size of an average atom. The STM has provided the first-ever images of individual atoms.

semicondutor   Naturally occurring semiconductors are those substances whose ability to conduct electrical currents lies between that of a good conductor and an insulator. These materials include silicon, germanium, and gallium arsenide. In almost all instances, however, the word semiconductor refers to artificial semiconductors, the so-called doped semiconductors, fabricated from natural semiconductors by introducing controlled amounts of impurities.

strong nuclear force   The strongest of the four forces of nature, it is the force between the color charges of quarks, and for this reason is sometimes called the color force or the chromo force, as in *quantum chromodynamics*. The nuclear force between protons and neutrons is then a "molecular" force arising out of the color forces among individual quarks making up protons and neutrons.

supercomputer A fast computer that sometimes uses a technology called *parallel processing*, in which a job is automatically divided up and the parts are processed concurrently by a set of central processing units.

superconductor At very low temperatures, below the boiling point of liquid helium, several metals and other alloys suddenly lose all natural resistance to the flow of electric currents through them. This property, sometimes described as a new state of matter, is called *superconductivity*. Electric currents can thus be transmitted without loss of power through a superconductor. Recently discovered *high-temperature superconductors* are an entirely different breed of materials, a copper oxide family of ceramics, which are poor conductors of electricity under normal conditions.

transistor A very reliable, versatile, and inexpensive switching device made from doped semiconductors. The invention of the transistor in 1947 heralded the high-technology age of information we know today. (See *semiconductor*.)

tunneling An exclusively quantum-mechanical effect in which a particle such as an electron seems to penetrate or tunnel through an otherwise inpenetrable force barrier, hence its name. Completely inexplicable by all the laws of Newtonian physics, such a bizarre event takes place because a particle sometimes behaves as a wave. It is like a ping-pong ball going through a steel door a few feet thick!

weak nuclear force Despite its name, this force is an entirely different force from the strong nuclear force. It is a newly discovered force, a very weak one, one of the four fundamental forces in nature, in which all the quarks and leptons interact with each other as equal participants. It remains one of the least understood forces.

wave-particle duality The defining foundation of quantum mechanics, that matter and radiation alike, within the reduced dimensions of the quantum world, behave at times like a wave and at other times, just as convincingly, like a particle. This is the most perplexing and paradoxical of all the mysteries of quantum physics.

Bibliography

Quantum mechanics

John Gribbin, *In Search of Schrödinger's Cat*. Bantam Books, 1984.

J.C. Polkinghorne, *The Quantum World*. Princeton University Press, 1984.

A.P. French and E.F. Taylor, *An Introduction to Quantum Physics*, The MIT Introductory Physics Series. Norton, 1978.

Quantum mechanics and high technology

T. Hey and P. Walters, *The Quantum Universe*. Cambridge University Press, 1987.

M.Y. Han, *The Secret Life of Quanta*. TAB Books/McGraw-Hill, 1990.

Quantum electrodynamics

R.P. Feynman, *QED*. Princeton University Press, 1985.

Jeremy Bernstein, *The Tenth Dimension*. McGraw-Hill, 1989.

Index